Information Society and its Manifestations: Economy, Politics, Culture

Borut Rončević / Matevž Tomšič (eds.)

Information Society and its Manifestations: Economy, Politics, Culture

PL ACADEMIC RESEARCH

Bibliographic Information published by the Deutsche Nationalbibliothek
The Deutsche Nationalbibliothek lists this publication in the Deutsche Nationalbibliografie; detailed bibliographic data is available in the internet at http://dnb.d-nb.de.

Library of Congress Cataloging-in-Publication Data
Names: Rončević, Borut, editor. | Tomšič, Matevž, editor.
Title: Information society and its manifestations : economy, politics, culture / Borut Rončević, Matevž Tomšič (eds.).
Description: Frankfurt am Main ; New York : Peter Lang, [2017]
Identifiers: LCCN 2016045742 | ISBN 9783631703557
Subjects: LCSH: Information society. | Information technology–Economic aspects. | Information technology–Political aspects. | Information technology–Social aspects.
Classification: LCC HM851 .I53187 2016 | DDC 302.23–dc23
LC record available at https://lccn.loc.gov/2016045742

Cover image by Ana Čefarin.

ISBN 978-3-631-70355-7 (Print)
E-ISBN 978-3-631-70356-4 (E-PDF)
E-ISBN 978-3-631-70357-1 (EPUB)
E-ISBN 978-3-631-70358-8 (MOBI)
DOI 10.3726/b10694

PL Academic Research is an Imprint of Peter Lang GmbH.

Peter Lang – Frankfurt am Main · Bern · Bruxelles · New York · Oxford · Warszawa · Wien

This publication has been peer reviewed.

www.peterlang.com

Table of Contents

Preface

Information societies are characterised by a high level of structural complexity. This causes significant challenges in societal steering and demands development of contextualised meta-governance mechanisms. In this volume we do not provide a detailed overview of a variety of conceptualizations of information society and related or partially overlapping concepts such as post-industrial society, post-Fordism, network society, post-modernity, knowledge society, etc. Instead, we provide insight into the bricolage of specific manifestations of information society in techno-economic, political and cultural spheres, unveiling the wealth of context-specific elements that have to be taken into account in analytical and practical work.

Borut Rončević and Matevž Tomšič
School of Advanced Social Studies in Nova Gorica
Gregorčičeva 19, 5000 Nova Gorica, Slovenia
borut.roncevic, matevz.tomsic@fuds.si

Perspectives of Information Society: Bricolage of Manifestations

Abstract: *Information societies are characterised by a high level of structural complexity. The processes of comprehensive societal change known as modernisation led to unprecedented dynamics and heterogeneity of institutional forms and practices. Increase in speed of symbolic communication, in which information technologies played the key role, further contributed to this process. In this introductory chapter we are dealing with the question of whether strategic steering of an information society is even possible and what are the consequences of differentiation, networkisation and informatization for steering and governance. We then present the overview of contributions to this volume, which offer insight into the manifestations of an information society in spheres of economy, politics and culture.*

Keywords: *information society, networkization, strategic steering, governance*

1. Introduction: Information in Complex Post-industrial Societies

Information societies are characterised by a high level of structural complexity. The processes of comprehensive societal change known as modernisation lead to unprecedented dynamics and heterogeneity of institutional forms and practices. Unprecedented speed and efficiency of symbolic communication, in which information technologies played the key role, further contributed to functional differentiation where particular systemic functions are performed by different subsystems with their specialised institutions and specific logics of functioning. Niklas Luhmann's social systems theory states that in contemporary society there is no central subsystem that would represent society as a whole (Luhmann 1995). According to Luhmann, different subsystems are completely autonomous and function on the basis of their special and particular interests and rationalities and each of them has its own medium of communication. Other authors have softened this claim by emphasising the reflection and systems discourse (e.g. Willke 1993), but there is a consensus about the de-centred nature of contemporary societies. In a similar way, for example, one of the key theorists of information society Daniel

Bell (1996) claims that in (postmodern) societies there is no unified referential system, no unified internal dynamic of structural change and no unified or dominant regulatory mechanism. They are composed of three different spheres, with each of them binding to its own 'axis-principle' that has an intrinsic status. These are: the techno-economic sphere that works on the principle of functional rationality; the political sphere that is based on the principle of legitimacy; and the cultural sphere whose axis-principle is self-fulfilment of the individual, i.e. his/her existential realisation as a creative human being (ibid. 29).

Furthermore, Bell acknowledges three key aspects of the post-industrial society. The first is the data, information describing the empirical world. The second is the organisation of this data into meaningful systems. The third is knowledge, i.e. the use of information to make judgments and make appropriate decisions (Bell 1973). Although Bell was writing about the coming of the post-industrial society, by conceptualising information as a key organising principle of this society he effectively introduced the notion of the information society; a society in which collection, creation, assembling, reproducing and massive commodification of information becomes the key activity, a part of all aspects of social organisation, and thus plays a crucial role in steering societal development.

In this volume we do not provide a detailed overview of a variety of conceptualisations of the information society and related or partially overlapping concepts such as a post-industrial society, post-Fordism, network society, high modernity, post-modernity, knowledge society, etc. It is also not our goal to review crucial contributions such as those of Daniel Bell, Manuel Castells, Niklas Luhmann, Ulrich Beck or Alvin Toffler. We already have very good literature providing these, such as Webster (2014) or Mackay et al. (2002). Instead we provide analytical insight to specific manifestations of information society in three aforementioned spheres as defined by Bell, namely techno-economic, political and cultural, trying to unveil some of the aspects of information societies that make it possible to steer them purposefully.

The dilemma of the development of societies should be subjected to planned steering or left to self-regulation, i.e. social evolution, is not limited to information societies. It is without doubt one of the most important and the most prominent dilemmas in the history of social, political and economic thought. From the reflexion of this issue arises the famous remark of Adam Smith (Wealth of Nations 1776): "People of the same trade seldom meet together, even for merriment and diversion, but the conversation ends in a conspiracy against the public, or in some contrivance to raise prices". We could add that the 'conspirators against the public' would have to meet to exchange the information required to coordinate their actions.

But the universality of specific developmental issues or questions does not presuppose the universal validity of solutions or answers. Conclusions based on a specific situation can be completely irrelevant to another. Therefore, while solving this dilemma in the context of information society, we have to approach it by considering the specific empirical level, focusing on structural characteristics of a defined social environment and on analysis of conditions that could enable or prevent specific ways of steering or self-steering social development. There is no universal answer to the above-mentioned dilemma and specific solutions work in specific societal constellations. This is the reason why attempts of transplantation of institutional arrangements that are successful in a specific environment, e.g. the highly successful information economy of the Silicon Valley, completely fail to meet the high promises and expectations in another, e.g. the planned 'Russian Silicon Valley' in Skolkovo near Moscow.

Is strategic steering of an information society even possible? Successful strategies have to take into account complex relationships and co-dependencies between various partial systems if they wish to achieve long-term success. This requires substantial hard and soft infrastructure, and significant effort at a structured exchange of information. It is therefore not surprising that the importance of regional innovation systems and other mechanisms for intermediation between relevant actors came to the fore with the advent of the information society.

2. From Differentiation to 'Networkization' and Coordination

Functional differentiation implies upgrading the effectiveness of a system since it enables the simultaneous steering of several processes that cannot be comprehended with a single context of conduct. Individual subsystems are very complex themselves since the processes of internal specialisation lead to the diversification of issues, principles and practices, highlighting a need for the reintegration of their diversified unity. Likewise, there is a need for reintegration of the system as a whole, to achieve systemic rationality. Furthermore, relations are determined by the diverging interests of key actors deriving from the very principles of functioning (for example, in a market economy whose profit-oriented logic triggers conflicts of interest between owners and employees) which, again, points out the need for co-ordination and mediation between stakeholders.

The development of modern information society is thus directed towards the independence of its different partial systems that are becoming "operatively closed", i.e. steered internally, on the basis of their own specific norms and principles. However, functional differentiation in terms of increasing specialisations in particular fields is coupled with the interdependence of these functions since every functional field can constitute a functioning society only when it is connected to the others (Willke 1993, p. 44). Since we are dealing with an increasing number of nodes at the macro (subsystems), meso (intermediary structures) and micro levels (individual organisations or actors) that communicate with each other with increasing speed and efficiency, we are also dealing with an increasing number of increasingly complex networks.

"Networkisation" of contemporary societies can thus be explained as inevitable consequence of modernisation, which, more than in the case of segmentary and stratification differentiation, put modern societies before the challenge of societal integration. These processes also led to the lessening of the probability of successful hierarchical coordination and successfulness of spontaneous evolution. Networks can offer part of the solution to the problems of social coordination that originate from developmental dynamics of modern societies. Transformational processes that accompany social modernisation led to the strategic process of formation of network forms. Messner (1997) offered a concise description of networkization – he did not use this expression to describe the trends – already twenty years ago.

One major trend is *movement in the direction of the organising society*. This implies that although numbers of collective actors and acceleration of developmental dynamics are growing, there is a reflection and awareness of the need to influence and steer societal development. Domination of nature and particularly of society, *Weltbeherrschung*, remains the major goal (Genov 1997, p. 412), also in the information society, although the instruments have to change. Genov terms this trend the 'spread of instrumental activism' and describes it one of the four main trends (1997, 2010). Advantages, derived from linking are also pooling (limited) resources and combining various competences and forming new, emergent ones that exceed the sum of resources of individual actors. Information technology only increased the relevance of this trend with the more and more efficient multi-level communication and the newly emerging ad hoc and virtual organisations, societies and communities.

Another major trend is the *increasing sectorization of economy and society* (Messner 1997, p. 150–153). This describes the process that we named functional differentiation. This process is complemented with the process of specialisation. This leads to the establishment of conditions for the increasing importance of individual partial systems or individual actors in society. Different sectors and actors are included into the process of decision-making, which is leading to the *supersaturation of the policy process.*

Differentiation of partial systems also leads to *the growth of policy*. It means that because of the risks, emanating from the prospect of decline of systemic rationality on account of target rationality, it comes to the increasing volume of state interventions (there are more and more domains and possible situations that have to be regulated) or the interventions and regulations by the supranational entities (European Union (EU), World Trade Organisation (WTO), etc.). Even public information exchange services, once almost synonymous with openness and unregulated accessibility, are now increasingly subject to regulation by national or supranational organisations, see for example the Great Firewall of China or debates on Internet neutrality in the EU.

These trends have immense importance for the state. In the long run this process led to the *decentralization and fragmentation of the state*. It means that various forms of intervention and steering started to emerge in the framework of the state. It also started shifting the responsibility on lower levels (regional, local). Furthermore, with decentralization came various forms of cooperation between the state and other actors. The *border between policy-makers and recipients of policies became blurred* (Ibid.). Jessop expanded the definition of the state beyond the usual conception of a political and administrative apparatus in control of a monopoly of power within clearly defined geographical boundaries and on the basis of work by Gramsci and Poulantzas develop the strategic-relational approach, the concept of the state as a social relation (Jessop 2007). This would explain why in some domains development led to the state – understood in the traditional sense – almost completely lost its abilities for autonomous policy-making. Input from non-state actors cannot be neglected without consequences. It came to *the inward loss of the autonomy of the state* (Messner 1997) in the relation to other partial systems. Globalization, the regionalization of economy and integration of national states into supra-national organisations contribute to *the loss of the autonomy of the state outwards.*

From this arose the need for a *cooperative or negotiating state* that has to learn to impart responsibility to other levels and other actors and of course to support the capacities of these actors for successful cooperation in policy processes (Ibid.). Conditions for the emergence of an *active society* (Etzioni 1968), in which strategic processes unroll in interaction between relevant actors and partial systems, are established on the basis of previous trends. This is at the same time the only successful solution to growing mutual dependency. These interactions establish relations of inter-systemic coordination which differ from market or hierarchical relations.

Different explanations of formation of new forms of social coordination also exist. The most famous of them is the approach of technological determinism, which Manuel Castells explained at the beginning of his most well-known work *The Rise of the Network Society*: "A technological revolution, centered around information technologies, began to reshape, at accelerated pace, the material basis of society. Economies throughout the world have become globally interdependent, introducing a new form of relationship between economy, state and society, in a system of variable geometry" (Castells 1996, p. 1). In this sense, for example, Castells also explained the downfall of socialistic systems as a consequence of their inability to adapt to demands that the revolution of information technology put before them (Castells 1997).

3. The State in the Network Society: Primus Inter Pares

In the beginning of the 1990s David Held famously stated that today we deal with a hybrid system, in which on the one hand despite different trends the system of sovereign national states still persists, but on the other hand systems of plural structures are also developing (Held 1991). The implication is that the national state represents only one of the existing centres of power in a broader network, where it often confronts other centres that limit its autonomy (Castells 1998, p. 304). However, this doesn't mean the inevitable decline of national state, but it indicates the changed role of the state in strategic processes: ... while global capitalism flourishes and national ideologies throughout the world explode, it seems that national state, created in modern era, loses its power, but, and this is essential, *not also its influence*" (Castells 1997, p. 243).

As we have already stressed, in a network society the state doesn't lose its autonomy only outwards, but also inwards. Castells states that subordinate social groups gain access to policy processes, especially in the lower levels. "Thus, a complex geometry emerges in the relationship between the state, social classes, social groups, and identities present in civil society" (Castells 1996, p. 271). In this way, lower levels, the so called "local state", become important strategic instances (Warner 1999). In this way, local and regional governments become a manifestation of decentralized political power, a point of contact between the state and other social subsystems. In this way, networks, within which it comes to policy-making, become much more complicated. Examples of this are the policies of European Union with the principle of subsidiarity, which complicates the analysis of networks with the introduction of an analysis of multi-level governance.

In view of the changing role of the state, a body of work emerged, attempting to introduce simplifying visions, which assign the guilt for the irrelevance of the state to economic globalization and the rise of the power of multi-corporations (see e.g. Martin and Schumann 1997). However, an even larger body of work about "localization" of competitive advantages indicates that the role of national states really is changing, but it is not becoming irrelevant. On the contrary, different levels are assuming completely new and different roles internally, with changing relations between national, regional and local levels as well as towards other actors. Instead of traditional macroeconomic measures that are necessary yet an insufficient condition for economic competitiveness, other factors, which can only develop in cooperation of various actors, come forward.

Weiss argued that the reaction of the state to pressures of functional differentiation wasn't uniform. In regards to political-institutional differences two answers formed. In both cases it is about the upgrade of coalitions: "upwards", in the direction of the construction of interstate coalitions on a regional and broader level, and "downwards", in the direction of the formation of coalitions with internal actors, for example in the form of a connection with the economy. She termed states that avail themselves of building these coalitions, "catalytic states". This means that in reaching goals, these states don't rely predominantly on their own resources – this is what "integral states" do. Instead they try to reach them as a central or dominant partner in the frame of coalitions of states, transnational institutions or private actors, where they try to be an indispensable link of particular strategic coalitions and at the same time they try to stay relatively independent in relation to other actors (Weiss 1998).

A national state in an information society actually finds itself under pressure from very different actors, spanning from capital and production networks to supranational structures and organised crime (Castells 1996, p. 304). However, it doesn't mean that those actors successfully exercise their influence or strengthen their autonomy. The shrinking autonomy of the state doesn't imply the automatic strengthening of the autonomy of other actors. One such example are post-socialist societies, where actors had insufficient resources and consequently were not able to make a substantial contribution to systemic transformation, with the consequence that post-socialist states actually had to increase their powers (Nielsen et al., 1995, p. 11) to be able to reduce them later. For more successful post-socialist countries, mainly those from East-Central Europe, this process took place during the accession to the EU, where significant efforts had to be invested in increasing administrative and judicial capacities of the state, the successfulness of which is still somewhat unclear (see Berend and Bugarič 2015).

The rise of the information society therefore introduced important changes not only to the role of the state, but also to scholarly debates on governance. Although the idea of 'governance without government' that is delivered by networks arose from the economic theory of the firm with roots in the 1930s and elaborated in the late 1980s, the ideas of heterarchical non-market forms of governance became especially prominent with the spread of information and communication technologies (Fukuyama 2015). Sometimes it seems that the notion of governing, defined as something that governments do (Howlett and Ramesh 2014, p. 318) is being pushed out of focus by governance, which is "about establishing, promoting and supporting a specific type of relationship between governmental and non-governmental actors in the governing practices" (Ibid.) Steurer, for example, models governance as Venn diagrams of three overlapping circles, each representing government (hierarchies), business (markets) and civil society (networks), which interact, hybridize and orchestrate, giving rise to the fourth mode of governance, the meta-governance (Steurer 2013). Although prone to failure, like other forms of governance, meta-governance allows for re-articulating and 'collaborating' among different modes of governance (Jessop 2009, p. 93) and seems most appropriate for societal steering in the information society.

4. Observing the Phenomena: Overview of Volume

In this volume, we present manifestations of an information society in the realms of economy, politics and culture. They represent the three main areas of societal life. The economy is a set of institutions, actors and activities oriented toward 'provision' of society with material and other resources that represent the basis of its existence and future development. Politics is a set of relations among public-oriented actors and their activities that result in the creation and implementation of collectively binding decisions; furthermore, it represents a space for the manifestation of different ideas, beliefs and interests as well as interactions between their agents in the decision-making process. Culture is the set of values, ideas, beliefs, and sentiments which give order and meaning to people's lives and which provide the underlying assumptions and rules that govern human behaviour in different spheres of society. We are speaking about the particular distribution of patterns of orientations toward objects, persons and phenomena among members of society. These three realms are autonomous and interrelated/interdependent at the same time. We can discern both similarities and differences in terms of relevance and use of information technologies.

Technological developments brought the rise of an embedded information economy, with all opportunities and constraints, related to it. As stated by

Granovetter (1985), economic rationality is embedded in social relationships, i.e. networks of mutual connections of individuals. Knowledge of the economic actors is generated and norms of their conduct are reproduced through their interactions, since they steer their actions on the basis of past experience from cooperation with others.

Utilisation of information developments is highly relevant for the strategic steering of economy. It is related to the ability of strategic conduct. As explained by Borut Rončević and Urška Fric, the strategic orientation of key actors is crucial for path-creation in terms of launching effective policies that would result in a developmental break through. This was the case with successful technological late comers like Ireland and Finland. It was the systemic discourse that changed the developmental trajectories of the two countries.

For the successful functioning of an information economy, proper normative infrastructure is necessary. Dolores Modic analyses the importance of intellectual property rights in the information sector, claiming that a sociological approach, using innovation systems theories, can be very insightful in this respect.

The rise of the information society resulted in the development of new approaches regarding cooperation in the field of economy. One of them, industrial symbiosis, is discussed by Erika Uršič and Borut Rončević. Industrial symbiotic networks connect various systems, including the social system. The authors present an analytical approach that falls within the realm of computational sociology and will model the morphogenesis of a complex network of networks that consists of technological, informational, logistical and social component.

The information society and the new economy influence the conduct of key decision-makers. The networking of the business elite is discussed by Lea Prijon. Its members merge in interlocking directorates, i.e. networks that provide access to relevant business-related information for important business transactions that result in significant economic and non-economic profits.

There are few fields where informantion management is more relevant than advertising. Jana Suklan and Vesna Žabkar present a synergy model with insights into how different components interact over time, applying a statistical modelling technique primarily used in financial markets to model advertising effectiveness.

Development of information-communication technologies puts some key political issues and challenges in a new perspective. It has an impact on political activities, mode of organisation and decision-making as well as on the very character of democratic polity. The innovative forms of political activity in the terms of e-participation, based on utilisation of information-communication technology, could stimulate political and civil engagement.

According to the objectives of democratic functioning there is a broad range of online tools trying to foster transparent and inclusive democratic participation. Matevž Tomšič and Petra Kleindienst thematise issues of e-participation and e-democracy using the examples of countries in Central and Eastern Europe. They present EUvox as an e-participative tool which was designed to increase the political competence of the citizenry, enabling them to make more informed political choices.

Trust is considered as an important factor of political participation. However, as Victor Cepoi argues, there are different types of trust, with diverging effects on citizen's engagement. His data shows that only generalized trust is relevant in this regard while political trust does not give any significant outcomes. New sources of information are significant in explaining generalized trust.

Many observers argue that experts play an increasingly significant role in decision-making processes in contemporary polities. Matej Makarovič and Mateja Rek claim that basing key decisions on knowledge and expertise is a necessary reaction to the increased complexity and the massive inflows of information as the key feature of the information society. However, they stress that some kind of 'rule of experts' cannot replace democratically elected political institutions.

New technologies provide opportunities for not only more effective, but also more inclusive problem-solving. Janja Mikulan Kildi presents the case study of the EUROSHA project in Kenya that was aiming at crisis preparedness by increasing the level of information of local and international humanitarian actors through open-source tools and information-communication technologies.

The change that is being brought by the 'informatisation' of society strongly affects the sphere of culture. We are speaking about cultural individualisation which results in the fluidisation of existing modes of identification. Some see these changes as 'liberation' of individuals from traditional (collective) bonds while other see them as generating a 'crisis of meaning'.

The information society refers to recent transformations of the social reality expressed in the form of semantic structures. Tea Golob emphasises that the excessive amount of information does not necessarily indicate that all people have the same access to it or that they equally able to make use of it. Individuals have become more reflexive and less bounded with social structural determinants, forming multi-layered, instrumental identifications which are not necessarily connected to their primary role-sets.

Information technologies have an increasing role in the education process. Jana Krivec identifies chunks in a human's procedural knowledge and describes their characteristics based on the reconstruction of chess variants. She presents a model that precisely defines the quantitative limitation of the chunks in our memory storage and process enhanced with the recombination mechanism of the elements in short term memory.

Effects of information society are evident also at the community level. Jernej Agrež and Nadja Damij deal with the information, learning and knowledge exchange in relation to natural disaster management, focusing on floods and the corresponding state of emergency in Lower Sava Valley. They stress the importance of a stable microscale information society network, present in the scope of local flood response activities.

Developments of new information-communication technologies are relevant even for the functioning of ethnic minorities. Dejan Valentinčič examines the impact of the social network Facebook on Slovenian minority and diaspora communities. His research results show clearly that young Slovenians from all nine countries use Facebook also as a tool for expressing their ethnic identity.

Our chapters discuss only a limited number of manifestations of an information society in the fields of economy, politics and culture. However, this bricolage provides highly relevant insight into the complexity of the processes, interrelations and interdependencies, induced by the development of information-communication technologies and offers some preliminary answers and points to opportunities for further exploration.

The volume shows a series of relatively independent phenomena that development of an information technologies poses a serious challenge for organisations of different fields of society. This is not to say that information matters while it didn't in the past. On contrary, the possession of relevant information (in military, business or other terms) was important in every period of human history. Specificity of current times in this regard refers to the management of information processes. Due to the abundance of information produced by and circulated through a number of different sources, it is often very hard to distinguish right from wrong, relevant from irrelevant, truth from manipulation. In order to do this in an effective way, the capability of information selection, interpretation of its essence and detection of possible flaws is of utmost importance. We are speaking about processing information. As stated by Nikolai Genov in this volume, its relevance comes largely from the need of organisations to strengthen the cognitive basis of their decisions and actions in order to increase the practical efficiency of the outcomes of their activities. Information resources can be perceived as the mechanism of upgrading organisational rationality. However, the process of upgrading organizational rationality is inherently controversial since it involves different pathologies as its by-products that might in certain historical situations even become dominant.

5. References

Bell, Daniel: "*The Coming of Post-Industrial Society.*" New York: Basic Books, 1973.

Bell, Daniel: "*The Cultural Contradictions of Capitalism.*" New York: Basic Books, 1996.

Berend, Ivan / Bugarič, Bojan: "Unfinished Europe: transition from communism to democracy in Central and Eastern Europe." *Journal of Contemporary History*, ISSN 0022-0094, 50(4), 2015 pp. 768–785.

Castells, Manuel: "*The Rise of the Network Society, The Information Age: Economy, Society, Culture*". *Vol. I.* Cambridge, MA: Oxford, UK: Blackwell 1996.

Castells, Manuel: "*The Power of Identity, The Information Age: Economy, Society and Culture*". *Vol. II.* Cambridge, MA: Oxford, UK: Blackwell, 1996.

Fukuyama, Francis: "Governance: What Do We Know, and How Do We Know It?" *Annual Review of Political Science*, 19, 2015, pp. 6.1–6.17.

Genov, Nikolai: "Four Global Trends: Rise and Limitations". *International Sociology*, 12(4), 1997, pp. 409–428.

Genov, Nikolai: "*Global Trends in Eastern Europe*". London and New York: Routledge, 2010.

Granovetter, Mark: "Economic Action and Social Structure: the Problem of Embeddedness". *American Journal of Sociology*, 91(3), 1985, pp. 481–510.

Howlett Michael / Ramesh, M.: "The two orders of governance failure: Design mismatches and policy capacity issues in modern governance". Policy and Society, 33(4), 2014, pp. 317–327.

Jessop, Bob: "*State Power: A Strategic-Relational Approach*". Cambridge: Polity, 2007.

Jessop, Bob: "From Governance to Governance Failure and from Multi-level Governance to Multi-scalar Meta-Governance". In: Arts, Bas; Langedijk, Arnoud; van Houtum, Henk J. (eds.): *The Disoriented State: Shifts in Governmentality, Territoriality and Governance.* Springer, 2009, pp. 79–98.

Luhmann, Niklas: "*Social Systems*". Stanford: Stanford University Press, 1995.

Mackay, Hugh / Maples, Wendy / Reynolds, Paul: "*Investigating Information Society*". London, New York: Routledge, 2002.

Martin, Hans-Peter / Schumann, Harald: "*The Global Trap: Globalization and the Assault on Democracy and Prosperity*". Leichardt: Pluto Press, 1997.

Messner, Dirk: "*The Network Society: Economic Development and International Competitiveness as Problems of Social Governance*". London, Portland: Frank Cass, 1997.

Steurer, Reinhardt: "Disentangling Governance: A Synoptic View of Regulation by Government, Business and Civil Society". *Policy Science,* 46(4), 2013, pp. 387–410.

Webster, Frank: "Theories of the Information Society". New York and London: Routledge.

Weiss, Linda: "*The Myth of the Powerless State*". Ithaca, New York: Cornell University Press, 1998.

Willke, Helmut: "*Sistemska teorija razvitih družb* [System-theory of developed societies]". Ljubljana: Fakulteta za družbene vede, 1993.

Nikolai Genov
School of Advanced Social Studies in Nova Gorica
Gregorčičeva 19, 5000 Nova Gorica, Slovenia
nikolai.genov@fuds.si

Information Resources in Upgrading Organizational Rationality

Abstract: *Max Weber interpreted the achievements of Western Europe as intrinsically related to accelerated social rationalization. He identified the core of the process in the purposeful collection and practical use of precisely tested and systematically organized information (knowledge). George Ritzer developed and applied these ideas in his analysis of modern society by focusing on the efficiency, calculability, predictability and control determining the success of the McDonalds fast food restaurants. However, Ritzer falsely presented Weber's ideas as theory while Weber aimed at developing sensitizing concepts for the understanding of a specific historical situation. Ritzer's stress on the achievements of rationalization understood as McDonaldization neglects the pioneering innovations in assembly lines copied in the fast food restaurants. He underestimates their tendency of McDonaldization towards organizational stagnation while the major task of the current strategies of organizational rationalization is change management. The discussion opens the vista to new conceptualizations of rationality types. They are related to stages in problem-solving activities leading to the upgrading of the rationality of organizations. The pairs of rationality types elaborated in the article include the systemic and sub-systemic, individual and collective, ultimate and instrumental, of the process and of the results, short term and long-term rationalities. The "hidden agenda" of the analysis is the linking of each of these types of bounded rationality to deviations from rationality (irrationalities). The analysis of organizational development shows the intrinsic link of the rationalities and irrationalities to the quantity and quality of the available and used information in organizational problem solving activities.*

Keywords: *social rationalization, bounded rationality, types of rationality, organizational irrationality*

1. Ideas in Macro-social Innovations

The rise of the industrial society was based on the expansion of new production technologies. They incorporated the advancements in rational building and empirical testing of theories in the sciences of nature. The theories increasingly absorbed and ordered various types of information and transformed it in systematic verified knowledge. It had the potential to support precise descriptions, convincing explanations and reliable forecasting of processes in nature – the cultivated nature including. On this cognitive basis recommendations for action with the high level of practical efficiency became possible. Step by step the spread of increasingly sophisticated technologies affected activities rather different from the industry. The influence of rationalized science radically changed the logistics of trade and transportation, the organization of public politics, the government of states and reached action spheres like architecture and music. In this way the influence of the industrial revolution together with the impacts of the (West) European Reformation and the Enlightenment fundamentally upgraded the quality of the purposeful collection, processing and use of information (knowledge). The task became the common ingredient of all major activities in this part of the world. By transfer of information and practices as well as by autonomous developments societies in other continents followed similar paths of development.

The emergence of the social sciences was conditioned by the same profound social change. The social sciences were expected to meet public expectations by offering precise descriptions and explanations awkward phrasing. Max Weber timely identified the urgent requirement of this historical moment and offered his explanatory vision of social rationalization. For him it consisted of the accelerated incorporation of rationalized knowledge (information) in the guidance of human action. He understood this process as the core of the ongoing social change in his times and as the major social process which will have its full development in the future.

Taking Weber's classical ideas as a starting point, George Ritzer attracted broad attention with his interpretation of the more recent phenomena of social rationalization. He took the success of the McDonalds fast food restaurants under close scrutiny. The outcome is the attractive metaphor of the McDonaldization of contemporary societies due to the increasingly important role of knowledge (information) in social development, in the changes of everyday life in particular.

These intellectual developments and their impacts on social change provoke a series of questions concerning the rise and the viability of the ideas of Weber and Ritzer as well as about the practical implications of the ideas. What are the real achievements of both sociologists in the intellectual efforts to make the social change transparent and manageable? What are the limitations of Weber's and Ritzer's ideas? Can we productively use some concepts and research strategies developed and applied by Weber and Ritzer in the continuing efforts to adequately describe, explain and forecast the ongoing social changes?

2. Weber and Ritzer: Achievements and Limitations in the Studies on Social Rationalization

The major achievement of Max Weber in the study of social change is his systematic stress on the relevance of ideas in the historical processes. In this way he substantially contributed to the better understanding of the emergence and worldwide spread of the capitalist economic and political structures. His interpretative approach to the Calvinist theological innovations fostered the complex analysis of the institutional changes during the (West) European Reformation and of its outcomes. Contrary to other influential views about the emergence and development of capitalist society Weber did not understand ideas (knowledge, information) as the emanation of economic and political situations. For him ideas are mutually related with material (practical) situations but have increasingly taken the lead in the shaping of social structures and processes after the Reformation. However, Weber's theoretical and methodological approach to the historical process was not exclusively focused on the role of ideas in it. In his concepts, research methodology and in his research he followed a broader understanding of the relationships between ideas and interests in social life. His strategic statement in this respect makes the nuances in his theoretical and methodological sensitivity and flexibility clear: "It is interests (material and ideal) and not ideas which have directly governed the actions of human beings. But the "worldviews" that have been created by ideas have very often, like switches, decided the lines on which the dynamic of interests has propelled behaviour" (Weber 2004, p. 69). It is exactly this last point concerning the strategic historical role of ideas (knowledge, information) which particularly shaped Max Weber's vision about the sources, content and effects of social rationalization.

The next key point in Weber's interpretation of the role of ideas in the historical turn to capitalist economic and political structures concerns the profound change in the content of the orientations and in the behaviour of the Europeans affected by the Reformation and Enlightenment. He analysed the predominant shaping of thinking and behaviour in the Western part of Europe by religion, traditions or emotions before the Reformation. He noticed than in the aftermath of the Reformation the previously dominating religious, traditionalist or affective shaping of thinking and behaviour became increasingly challenged and replaced by the knowledge-based (information guided) codifications of the everyday thinking and behaviour. The codification of thinking and behaviour made the rational calculation of advantages and disadvantages in the decision making possible. The same rationalizing effect shaped the behaviour of social actors, the predictability and the control of the human action. The rationally codified thinking and behaviour was incorporated in organizations for the rationalized handling of economic, political and cultural processes. The consequence was the high efficiency of the formalized bureaucratic organizations. It is this efficiency in handling social problems that makes the formalized organizations regular winners in the competition with all other forms of organization of social life. In the times of Weber as well as today the formal organizations were and remain the major actors shaping the future of social life.

Weber put this diagnosis on the social processes he witnessed and offered his explanation focusing on the changing ideas and the changing role of ideas in social life. He regarded the social rationalization as a triumph of the calculability, predictability and efficiency incorporated in the bureaucratic rules of the formal organizations. However, he was quick in noticing that the process was also full of controversial trends and effects. Therefore, his diagnosis of the social rationalization is marked by some pessimistic overtones. On various occasions Weber underlined his conclusion that the triumph of social rationalization is bitterly paid and will be paid by the individuals with the loss of personal creativity, individuality and affective life (Weber 2002, p. 975). Consequently, the spread of the information based rationality of formal organizations in modern societies should be analysed and evaluated as a multidimensional and deeply controversial process. It comes about in social history accompanied by many gains and many losses for formal organizations, for societies and for human beings (Whimster and Lash 2014).

George Ritzer took Weber's idea about the controversial processes and effects of social rationalization out of its typical academic context and made it popular among broader audiences. He did this by focusing his analysis on the guiding rationality of the business success of the McDonalds fast food chain. His metaphor about the McDonaldization of the global society became widely known due to the clarity and the close-to-the-earth argumentation of Ritzer. According to him, the core of the organizational philosophy and practice of the McDonaldization is reducible to four principles which summarize the rationality of organizations in modern societies best – efficiency, calculability, predictability and control. Repeating Weber, Ritzer underlines the point that these principles and the related practices of the rationalized formal organizations are sources of both rationalities and irrationalities. Therefore, the organizational rationalization implemented by the McDonaldization of the world has controversial features and brings about controversial consequences. Some of them are in accord with the constructive principles of organizational rationality and therefore deserve to be supported and strengthened. Others lead to dehumanizing and organizationally destructive effects. They should be moderated or eliminated as far as possible (Ritzer 2013).

The direct repetition of Weber's argumentation and Weber's terminology in Ritzer's book on the McDonaldization of the world is based on his essay (2008) "The Weberian Theory of Rationalization and the McDonaldization of Contemporary Society". This link and Ritzer's interpretation of Weber's ideas provoke critical and constructive reactions. The first point to raise concerns the very title of the essay. Has Weber really developed a *theory* of rationalization? The answer depends on the methodological understanding of the content and functions of a theory. Weber's point of view is very clear in this respect. He vehemently rejected the understanding of theory which dominated the natural sciences in his times and was shared by numerous influential sociologists from Emile Durkheim on. According to this understanding theory is a conceptual framework which makes repetitive and comparable descriptions, explanations and forecasting of comparable situations possible. Weber assumed that this type of universalized theory is impossible in the social sciences. Instead, in numerous methodological debates he presented the theory building as development of sensitising concepts to be used in the analysis and explanation of unique historical situations and processes. This strategy fully applied to his distinction of types of rationality to be used in the analysis of the social rationalization marking the emergence and development of the European modernity. Weber understood his task as limited to the explanation of a unique historical constellation of factors, processes and effects – and nothing more. No generalized theory of social rationalization was on his research agenda. Moreover, he understood the task of building a generalized theory of social rationalization as badly conceived and unrealistic.

Weber had good reasons to link the specific historical constellation of events bringing about social rationalization in the post-Reformation Europe as intrinsically related with the increasing relevance of formal organizations. Basically the formal organizations are collective actors. Therefore, one may expect that the issue of collective rationality should be a key issue in his efforts to explain the role of formal organizations in the rationalization process. Contrary to this expectation, there is no text by Weber in which he deals with issues of collective rationality in the context of social rationalization (Kalberg 1980). This omission is not by chance. Weber is very precise in his own definition of what his sociological studies could or should achieve and how. Their task is to understand actions related to other human beings (social actions). "Action" itself is defined by Weber as a phenomenon "insofar as the acting individual attaches a subjective meaning to his behaviour – be it overt or covert, omission or acquiescence" (Weber 2002, p. 4). This interpretative approach to action in general and social action in particular is clearly individualistic. This characteristic should not imply any underestimation of this specific approach to the tasks and methodological tools of sociology. The interpretative approach might be productive in the sociological studies on organizations and rationalization. However, Weber's definition of the tasks of sociology puts substantial limitations on the sociological studies. His interpretative sociology neglects the holistic parameters of organizations and organizational change and thus compromises the relevance of his approach to the upgrading of the rationality of organizations *as collective actors*.

Key ideas of Ritzer's analysis and argumentation are also questionable if they would be put under scrutiny from historical, methodological and substantive points of view.

First, Ritzer is certainly right in his identification of a social trend of increasing efficiency, calculability, predictability and control in our world dominated by formal organizations. His linking of this development to the growing organizational rationality and the concomitant organizational irrationalities in the McDponaldization of the world is also well taken (Ritzer 2013, pp. 14–18). However, his diagnosis turns out to be profoundly incorrect when put in the proper historical perspective. He neglects radical organizational innovations introduced in the Fordist assembly lines by applying the Taylorist stress on efficiency, calculability, predictability and control in the organization of mass production. The Fordist and Taylorist organizational innovations were copied and transferred to the 'assembly line' of the fast food McDonalds' restaurants *much later*.

Second, Ritzer's concepts of organizational rationality and rationalization don't take the multidimensionality of processes *in* and *of* modern organizations fully into account. A well differentiated and properly integrated concept of organizational rationality is needed in order to meet the requirements of the studies on organizational rationalization. Only well differentiated concept of rationality can meet the needs for practically relevant conclusions and recommendations which may efficiently guide the management of organizational change and the upgrading of the rationality of formal organizations.

Third, the crucial reason for the deficit of correspondence between Ritzer's ideas about the organizational rationalization is the limitation of the organizational pattern of McDonalds' fast food restaurants. Their organizational structures and functioning are focused on the achievement of a high level of efficiency, calculability, predictability and control *and on the stabilization of this achievement*. This is an outdated organizational strategy. The core strategy in present day organizational theory and practice is not the McDonaldization as presented, praised and criticized by Ritzer but a strategy which has an entirely different focus. It is the organizational strategy for the *management of change*. This is a radical reorientation from the stabilization and reproduction of established organizational structures and functions towards the permanent focus on the search and support to change and innovations *in* and *of* organizations as an active adaptation to changing environments. The organizational changes and particularly the changes bringing about upgrading of the organizational rationality are multidimensional as a rule. The management of change requires multidimensionality of the conceptual framework, of the study and the management of the upgrading of organizational rationality.

Therefore, two tasks have to be dealt with in parallel. *First*, a differentiated concept of rationality and complementary irrationality should be briefly developed. The concept will have systematically interconnected components by far exceeding the famous Weberian distinction of practical, theoretical, substantive and formal rationalities. This will be done by organizing the types of rationality and irrationality as pairs related to the stages of the problem solving activity. *Second*, the parallel task will be the demonstration of the relevance of each pair in descriptions, explanations and prognostications.

3. Variety of Rationalities and Irrationalities in Organizational Problem Solving Activities

The large body of publications on social rationality is very much focused on the *idea*, or, more precisely, on the variety of *ideas* related to the topic of social rationality and social rationalization. Any effort to systematically presenting these ideas discussed in influential publications (Hodgson 2012; Amoretti and Vassallo 2013) would go beyond the tasks of the present article. For the purposes of the following argumentation it would suffice to mention three widely recognized pillars of the discussions on the social rationality. They make it clear how narrow George Ritzer's views about social rationality and social rationalization are. *First*, the discussions on rationality are typically focused on or related to the conditions, content and consequences of the problem-solving *action* of individual and/or collective actors. Types of actors and stages in the implementation of activities are put under scrutiny. *Second*, the vast majority of the participants in the discussions on social rationality lays the stress on the *sound reasoning* in the cognitive preparation, guiding, control and evaluation of the individual and/or collective action aiming at problem resolution. *Third*, according to many but not all relevant authors the rationality of action is intrinsically bound to the *efficiency*, *effectiveness* or *optimality* of the action. The different approaches to these three pillars of the discussions on rationality have as a consequence the variety of concepts of social rationality which are in circulation. The list includes concepts of theoretical and practical, instrumental and substantive, individual, collective, short-term, long-term, etc., types of social rationality.

The crucial step towards bringing order in the concept(s) of rationality and social rationalization suggested here is the link between rationality types and the stages of problem-solving activity. More precisely, the suggested approach is closely connected with the studies on change management of formal organizations. The studies typically focus on the variety and the interconnectedness of the relevant parameters of rationality in the problem-solving organizational activity (Jabri 2012).

The starting point of these studies refers to the emergence of a *problem situation* as a rule. The problem situation is the outcome of the interplay of the organization under scrutiny with its environment. It consists of other individual and collective actors. The collective actors are mostly formal organizations. The goal oriented activities of interacting organizations might be cooperative, complementary or competitive. Competition between organizations is omnipresent and often fierce. Various organizations try to attain the same or similar goals under the conditions of scarce resources available for supporting their action. The typical assumption in this context is that organizations which are better informed, able to learn faster and have the abilities to apply their informational and material resources more efficiently would achieve better and possibly optimal results in the competition for achieving their goals. Poorly informed organizations which have difficulties in collecting, processing and using information, those possessing smaller informational and other resources and which are less able to use them efficiently are expected to fail or to achieve sub-optimal results in their goal attainments. In theoretical terms the organizational potentials, processes and results are seen as facilitating, contributing to or hindering the organizational rationality.

Therefore, the key parameters of the rationality of an organization are typically defined with a view to the *quantity and quality of the information (knowledge) a given organization possesses about the conditions, ends, means, processes and results of organizational action as well as to the quantitative and qualitative parameters of the action's outcomes.* The variety of factors influencing the rationality of a formal organization and the dynamics of its societal context are an indication that there are always possibilities for deviations in the problem-solving organizational activity towards irrationality. Simply presented, this means organizational development towards losing information or declining capacity to learn and therefore attainment of suboptimal or even self-destructive outcomes of the organizational activity. The corresponding meta-theoretical conclusion is therefore that *the rationality in the problem-solving activities of formal organizations is always bounded by facilitating or constraining circumstances and is therefore accompanied by irrationalities in the cognitive preparation and guidance of the problem-solving activities, in the problem-solving process and in its outcomes. They might not meet the requirements for optimal adaptation of the organization to the environment, might be not sufficient for its sustainable reproduction or even destructive for the organization* (Munro 2009; Spiegler 2014).

The so outlined complexity of the issues related to organizational rationality and irrationality makes the task for conceptual differentiations and synergy of the sub-types of rationality and irrationality rather complicated. But it is exactly for this reason that the task is quite urgent as well (Berg 2008). The disentangling (decomposition) of rationality/irrationality types can be implemented in several directions and at several structural levels of social action. Here the decomposition and the synergy of rationality types takes into account the multiple rationalities (and related irrationalities) along the line of the problem-solving activities of formal organizations. The strategy for reducing the perplexing over-complexity is the ordering of the types of rationality according to the stages in the organizational problem-solving activity. The next step is the organizing of the types of rationality in polar pairs by following Weber's innovative approach to the issue.

Thus, the starting point of the problem-solving activity of formal organizations is the emergence of a *problem situation* facing the organization under scrutiny. The problem situation is the outcome of the interplay of the organization with other individual and collective actors. The content of the activities to be implemented is determined by the technological, economic, political and cultural conditions which caused the problem situation. Provided there is the organizational ability and will on the part of the organization to start problem-solving activities, the next stage in the chain of activities is the *definition of the specific problem to be resolved by the organization.* If the governing instances of the organization really intend to resolve the problem situation and the organizational capacities would make this possible, then the next step in the chain of problem solving activities is the *decision(s) for action.* In the case that the resources and the abilities of the leadership of the organization offer enough potential for performance, what follows is the *implementation of the decision and control on the implementation.* Finally, organizations usually carry out *analysis and evaluation of the results* of the problem-solving activities thus preparing another round for problem-solving activities.

The chain of problem-solving activities is conditioned by the very character of formal organizations. They are goal-oriented systems which are able to change their own structure, functions and developmental paths in order to actively adapt to changing environments. This might happen in various ways and with various consequences. Therefore, in the outline of stages in the problem-solving activity some peaks of the intensity and relevance of pairs of organizational rationality or irrationality can be clearly identified. For instance, in the time span between the emergence of the problem situation and the definition of the problem to be solved two pairs of rationality/ irrationality predominate in the ongoing processes.

The *first* pair of rationality/ irrationality is related to the distinction between the formal organization as a systemic whole and its sub-systemic components. As a rule, both the systemic whole of the formal organization and its sub-systemic components (branches, departments, military units, etc.) are able to learn and guide the interactions between the components as well as between the components and the systemic whole of the organization. The components (sub-systems) of the organizational system might become differentiated or integrated in new patterns which might turn to be more or less efficient. New models of orientation, decision, coordination and action of organizational subsystems or of the organization as a whole could increase or decrease the quality of the problem-solving activity of the organization under scrutiny, etc. Therefore, the decomposition of the first pair of rationality types takes into account the interplay of sub-systemic components of the organization under scrutiny between themselves as well as with the systemic whole of the organization. General statements concerning these relationships, their structure and dynamics could be only tentatively formulated. The core issue concerns the relationships between the *systemic* rationality of the organization and the *sub-systemic* rationalities of its components. These relationships might be rather different in hierarchical organizations and in organizations with flat structures and a high level of decentralization in decision-making and control. The combinations of sub-systemic and systemic rationalities and irrationalities are related to the dynamics of interplay of the active agents since the organization as a systemic whole and its sub-systems are able to learn, act, interact and change in a relatively autonomous way (Maula 2006).

The linking of the problem situation with the efforts to define the problem which is to be resolved is the peak in the relevance of a *second* pair of rationality/ irrationality. The distinction of the pair and its polar parts is rooted in the double-sided characteristic of the vast majority of formal organizations as collectives and simultaneously as composed of acting individuals. The individuals occupy various organizational positions and implement specific roles in the organization. The typical present-day formal organization can be regarded as a collective of individuals too. Moreover, in the usual formal organization including hundreds or thousands of individuals there are small or larger collectives at various structural level of the organization. These collectives have specific interests and might act as collective actors under circumstances. Therefore, the preparation and implementation of a problem-solving activity by a present-day formal organization are marked by the commonalities, differences and mutual influences of *collective* and *individual* rationalities at various structural levels and in various contexts. The characterization of the individual rationality in organizations requires a precise definition of the organizational positions (status) and the roles of the individuals under scrutiny in the organizational context. Groups of individuals occupying similar or different positions and having similar or different interests in the organization might be thus identified (Denis 2002). Due to the commonalities, divergences or contradictions of group interests the action rationality of individuals or of collectives of individuals could support and strengthen the collective rationality of the organization or undermine and potentially destroy it (Weirich 2010). Therefore, the distinction and consideration of both types of organizational rationality as well as balancing them is one of the key tasks to be resolved in the efforts to upgrade the aggregated rationality of the organization under scrutiny (McGregor 2012).

If the governing instances of the organization would be interested in the problem solving and possess the organizational will and resources for the purpose, the efforts to define the problem to be resolved continue in a series of specifications. What kind of decisions should be taken, by whom and in what time sequence? Who should implement the decisions and how? The answers might be very much differentiated since the problem-solving activities of organizations are typically guided by a tree of ends. Some of them are related to the expected final results generally orienting the chain of problem-solving activities. They are the *ultimate ends* of action in the given context. Other ends are focused on the issue *how to achieve* the once defined ultimate ends. These *instrumental ends*, or goals, have the status of means to be used for the achievement of the ultimate ends of the problem-solving activities. In some situations the ultimate ends of specific problem-solving activity might be related to the ultimate values in human life like survival, truth or beauty but this linkage is not obligatory. In the vast majority of cases the ultimate ends of specific problem-solving activities are focused on more modest tasks. Some of them might concern the simple improvement of the links between functions of the organization (Tomer 2008). Nevertheless, the character and the quality of the balancing of *ultimate* and *instrumental* rationality incorporated in these two types of ends of problem-solving activity substantially determine the outcomes of the activity. It is a basic assumption of bounded rationality that there could be and usually are irrational deviations from the ultimate and instrumental rationality in every specific case of organizational problem-solving activity.

The implementation of the problem-solving activities raises the relevance of another pair of rationality/ irrationality. The content of the efforts to invent, select or adjust the means to the ends of problem-solving very much depend on the *predominant* orientation of the activity towards achieving cognitive results (information gathering and processing) or towards practical change of matter, energy, capital, political power, etc. These types of orientation and outcome of action characterize the distinction between the *cognitive* (Nelson 2008) and the *practical* (Sandberg and Tsoukas 2011) rationality. The cognitive rationality is strategically guided by the ideal of truth. Thus, the achievements and failures in the search for cognitive rationality are measured according to criteria for the development of concepts and statements containing truth about reality. The practical rationality is guided by the ideal of efficiency. Both types of rationality are closely interconnected. Typical cognitive aims might be achieved by applying mostly practical means as this often happens in scientific experiments. Vice versa, practical achievements in industrial technologies or in political coalition building might be achieved by exclusively relying on properly designed and efficiently used cognitive (intellectual) means. The link between truth and practical efficiency might be rather uncertain under circumstances, however. In some activities like politics or diplomacy practical efficiency (practical rationality) might be achieved on the basis of various degrees of intentional or not-intended deviation from the ideal of truth.

The definition of the problem to be resolved, the decisions concerning the ends and means of the problem-solving activity and the modalities of its implementation are inherently related to the *procedural* rationality (Laville 2000). Once the outcomes of the problem-solving activity are in sight, the major concerns of the actor (actors) promoting organizational change usually shift from the issues of the procedure to the *resultant* rationality (Simon 1995). This re-focusing of the analysis, control and corrections has retrospective and prospective dimensions. The desirable or expected results are evaluated with a view to the intended or already implemented change. However, in numerous specific cases the results of a chain of organizational problem-solving activity are regarded only as intermediary results in a continuing longer chain of problem-solving activities. In this latter case the results already achieved are being evaluated with a view to their potential relevance for the handling of future challenges.

All organizational problem-solving activities have temporary parameters. The decision to indulge into these type of activities – and not into entertainment, for instance, the implementation of organizational decisions to act in a specific way and the analysis of the results achieved might very much vary if short-term or long-term results are intended. Both types of action are guided and evaluated by two different types of rationality – the *short-term* and *long-term* rationality of action (Stephan 2010). Their combinations and balancing might come by chance. But in most cases of management of organizational change the combination and the balancing of both types of rationality is the subject of very careful considerations and calculations. Some organizations develop and apply the strategy of focusing on consecutive short-term portions of action. The background of this preference is usually the assumption that the repetition of successful short-term rationality in action might continue for long. The assumption might be correct but it might be also very much misleading and bring about organizational catastrophe. Other organizations are ready to allow short-term irrationalities (failures, losses) into their problem-solving activities. They follow the presumption that the expected long-term positive effects (rationalities) promise a higher level of sustainability for the organization under scrutiny. The so outlined conceptual framework might be applied in the analysis of the change management implemented by companies like "Nokia", by nation-states in dynamic transitional periods or by supranational organizations like the European Union in cases of crisis management (Genov 2012, pp. 6–10).

4. Conclusions

The above analysis and argumentation leads to four major conclusions. *First,* the time has come to try the development of a systematic and simultaneously well differentiated *concept* of social rationality. *Second,* the development of this concept is expected to be intrinsically bound at the concept of problem solving activities. *Third,* the cognitive consistency and instrumental efficiency of this concept could and should be tested in its applications. The simplest task in this respect is the use of the concept as guidance for systematic descriptions. Much more sophisticated tasks are related to the use of the concept as guidance for multidimensional explanations and prognostications. *Fourth,* the most promising field for testing the visions about multiple rationalities and irrationalities is the problem solving activity of formal organizations.

In this broad context special attention should be paid to the conditions, processes and effects of upgrading the rationality of organizations. The process is usually understood as rationalization of the organization under scrutiny. The relevance of the information resources in the efforts to achieve upgrading of the rationality of formal organizations is a key issue in the rationalization of organizations. The reason is the rapidly growing global relevance of the perception, storage, processing and use of information in the organizational problem solving activities. The impetus for the increasing relevance of the processing information comes largely from the need of organizations to strengthen the cognitive basis of their decisions and actions in order to increase the practical efficiency of the outcomes of their activities.

There is still another more general conclusion stemming from the above analysis and argumentation. It reads that the improvement of the cognitive basis of organizational activities and the increase in organizational efficiency characterize the global trend of *upgrading the rationality of organizations.* The trend is manifest and influential at all structural levels of social life, from person-to-person interactions in grass roots politics to the economic enterprises and the exchange in the global markets. The major structural mechanisms implementing this global trend are the differentiation and integration of organizational units. The involved agencies of individual and collective actors initiate and perform these processes on the basis of organizational learning and activity. Given the large variety of structural and actor-related factors influencing organizational processes, this trend is neither linear nor characterized by constructive effects alone. On the contrary, the upgrading of organizational rationality is usually accompanied by the emergence, spread and impacts of organizational pathologies. They might be brought about by both the differentiation and integration of structures and functions of organizations. This is the reason why the orientations, decisions and actions of organizational actors are usually the source of both organizational rationalities and irrationalities, or factors in their reproduction. Thus, the global trend of upgrading organizational rationality is inherently controversial. Under certain circumstances the organizational pathologies coming about as by-products of the upgrading of organizational rationality might come to dominate large-scale historical situations marked by the move of organizations towards the increase or decline of their organizational rationality.

5. References

Amoretti, Maria C. / Vassallo, Nicla (eds.) *Reason and Rationality*. De Gruyter: Berlin, 2013.

Berg, Anne M.: "Organizational Rationalities and Change Management". In: Berg, Anne M. / Eikeland, Olav (eds.) *Action Research and Organization Theory*. Peter Lang: Frankfurt am Main, 2008, pp. 65–81.

Denis, Andy M. P.: "Collective and Individual Rationality: Maynard Keynes Methodological Standpoint and Policy Prescription". In: Zarembka, Paul (ed.) *Confronting 9–11, Ideologies of Race, and Eminent Economists*. JAI Press: Amsterdam, 2002, pp. 187–215.

Genov, Nikolai: "The Challenge of Four Global Trends". In: Genov, Nikolai (eds.). *Global Trends and Regional Development*. Routledge: New York, 2012, pp. 1–24.

Hodgson, David H.: *Rationality*. Oxford University Press: New York, 2012.

Jabri, Muayyad: *Managing Organizational Change: Process, Social Construction and Dialogue*. Palgrave Macmillan: Basingstoke, 2012.

Kalberg, Stephen: "Max Weber's Types of Rationality: Cornerstones for the Analysis of Rationalization in History". *The American Journal of Sociology*, 85(5), 1980, pp. 1145–1179.

Laville, Frederic. 'Foundations of Procedural Rationality'. *Economics & Philosophy*, 16(1), 2000, pp. 117–138.

McGregor, Douglas: "Theory Y: The Integration of Individual and Organizational Goals". In: Godwin, Mary / Gittell, Jody J. (eds.), *Sociology of Organizations: Structures and Relationships*. SAGE: Los Angeles, CA, etc., 2012, pp. 248–258.

Maula, Marjatta: *Organizations as Learning Systems: "Living Composition" as an Enabling Infrastructure*. Elsevier: Boston, 2006.

Munro, Alistair: *Bounded Rationality and Public Policy: A Perspective from Behavioural Economics*. Springer: Dordrecht etc., 2009.

Nelson, Richard R.: "Bounded Rationality, Cognitive Maps and Trial and Error Learning". *Journal of Economic Behavior & Organization*, 67(1), 2008, pp. 78–89.

Ritzer, George: "The Weberian Theory of Rationalization and the McDonaldization of Contemporary Society". In: Kivisto, Peter (eds.). *Illuminating Social Life: Classical and Contemporary Theory Revisited*. Pine Forge Press: Los Angeles and London, 4th ed., 2008, pp. 41–59.

Ritzer, George: *The McDonaldization of Society*. Thousand Oaks: SAGE, 7th ed., 2013.

Sandberg, Jörgen / Tsoukas, Haridimos: "Grasping the Logic of Practice: Theorizing through Practical Rationality". *The Academy of Management Review*, 36(2), 2011, pp. 338–360.

Simon, Herbert A.: "Rationality as Process and as Product of Thought". In: Bell, David E. / Raiffa, Howard / Tversky, Amos (eds.) *Decision Making. Descriptive, Normative and Prescriptive Interactions*. Cambridge University Press: Cambridge etc., 1995, pp. 58–77.

Spiegler, Ran: *Bounded Rationality and Industrial Organization*. Oxford University Press: Oxford, 2014.

Stephan, Gunter: "Short-run and Long-run Adjustment to Environmental policy: A Neo-Austrian Approach". In: Faucheux, Sylvie / O'Conor, Martin / van der Straaten, Jan (eds.) *Sustainable Development: Concepts, Rationalities and Strategies*. Dordrecht etc.: Kluwer, 2010, pp. 117–140.

Tomer. John F.: "Beyond the Rationality of the Economic Man, toward the True Rationality of Human Man". *The Journal of Socio-Economics,* 37(5), 2008, pp. 1703–1712.

Weber, Max: "Max Weber's Introduction to the Sociology of Religion". In: Weber, Max. *The Protestant Etjhic and the Spirit of Capitalism.* Roxbury Publishing Company: Los Angeles, 1995[1920], pp. 13–31.

Weber, Max: "Introduction to the Economic Ethic of the World Religions". In: Sam Whimster (ed. and trans.). *The Essential Weber.* London: Routledge: London, 2004, pp. 55–80.

Weber, Max: *Economy and Society: An Outline of Interpretive Sociology.* University of California Press: Berkeley, 2002, 7th printing.

Weirich, Paul: *Collective Rationality. Equilibrium in Cooperative Games.* New York: Oxford University Press: New York, 2010.

Whimster, Sam / Lash, Scott (eds.): *Max Weber, Rationality and Modernity.* Taylor and Francis: Hoboken, 2014.

Embeddedness in the Information Economy: Opportunities and Constraints

Borut Rončević and Urška Fric
Faculty of Information Studies in Novo mesto
Ljubljanska cesta 31a, p. p. 603, 8000 Novo mesto, Slovenia
borut.roncevic, urska.fric@fis.unm.si

Path-creation in the Information Society

Abstract: *Although the development of successful technological latecomers is thoroughly analysed, it still leaves room for additional research, especially in the context of information societies. In this paper we analyse two successful latecomers in the information economy (Ireland and Finland) and one less successful (Slovenia). Their path-dependent development is analysed in three phases: diagnosis of their available choices and constraints; the critical strategic decision representing a strategic shift; and systemic discourse, which changed developmental trajectories of the two countries.*

Keywords: *path-dependency, information society, Ireland, Finland, Slovenia*

1. Introduction

The paper deals with societal communication as part of strategic processes of technological latecomers in their efforts to overcome their path-dependent relative developmental lag and join the ranks of the most competitive nations. These processes have already been well documented, but nevertheless require – in the authors' opinions – further scrutiny, with a special view on the specifics of the information society. The issues of societal communication, social steering or 'ideal' modes of governance have always been in the interest of socio-economic thought, while answering the question how development should be approached: either as a self-evolving phenomenon (societal evolution, market governance) without external interventions or as an area governed through top-down (hierarchical governance) or concerted efforts and stimuli (networked governance). The age old dilemma does not require a universal answer. Societal steering mechanisms are context-specific and should be addressed on a case-by-case basis – depending on the specific social and policy context at hand. After all, as Jessop (2008) points out, all modes of governance – even meta-governance, combining all other types of governance – are prone to dilemmas, contradictions, paradoxes, and ultimately, failures governance (Jessop 2008). This is even more relevant in the context of the information society, which brought unprecedented acceleration to social dynamics and an increase in complexity. Trying to implement and transfer uniform solutions within distinct milieus in the information society could prove questionable to say the least, as the outcomes of such measures and institutional arrangements would arguably imply a counter effect, leading to excessive costs and inefficiencies.

From the research perspective, analysis of disparate social settings and their specific policy guidelines is restricted by not bearing the same degree of relevance in different settings, thus limiting any generalizability and validity of the corresponding findings. Even more so also due to the fact that technological development in the information society is inherently multifaceted process that exhibits strong connections to other societal subsystems. Consequently, any sustainable strategies of technological development and related research have to consider complexities and co-dependencies arising from co-evolution and interactions within the societal constellation of information society as a whole.

2. The Theoretical Backgrounds of Path-dependency

David discusses path-dependency in the sense of how "events and decisions that have taken place in the past continue to influence current decisions and ways of working such that people become locked in paths that they cannot break free of" (in Morrey et al. 2012, pp. 3). In a similar line of reasoning Domanski (2003 pp. 134) interprets path-dependency as a two-fold concept: as interdependencies between choices made over time (in the broader sense) or when historical sequences of contingent events set into motion mechanisms that have deterministic properties (in the narrower sense).

Sydow et al. (2005 pp. 2–3) posits that the classical theory of "path-dependency" has been developed as an alternative to neoclassical economics. Paul David further developed this theory as a critical response to the study of QWERTY keyboard standard sustainability between 1985 and 1986 (Sydow et al. 2005, pp. 2–3). Despite the perceived technological inferiority, the standard succeeded in replacing other more sophisticated alternatives and achieved worldwide adoption. The inefficiencies of monopolies from the path-dependency perspective, which are key to understanding micro-economic relations, are thus illustrated by David (in Sydow et al. 2005, pp. 2–3) using two conjectures: (i) except for the most efficient solution market pressures render all alternatives unviable and (ii) in the majority of cases, all decisions are irreversible and can be undone only if a clearly better alternative emerges. Contrary to these conjectures, the theory of path-dependency introduced a different dynamic, where the present and future states are connected by past decisions and also influenced by random events (Sydow et al. 2005, pp. 2–3).

In 1989 the Economic Journal featured a study by Brian Arhur, who investigated a case of two competing technologies, where one of them took the adoption initiative not due to technological superiority, but due to several conditional events that were stochastically related to the initial conditions of the model (Greener 2004, pp. 4–5). The presented case study is also similar to the VHS recorder standard (Liebowitz et al. 1995, pp. 209) and demonstrates that even a single event or decision can lead to product lock-in, even though the chosen alternative has been proven as the less efficient and economic choice (Morrey et al. 2012, pp. 3).

Meyer and Schubert (2007, pp. 24) refer to David's technological interrelatedness and Arthur's competing technologies studies when defining path-dependency within the technological domain, however, the concept can be observed and appreciated also in the wider sociological sense, which positions technological paths as a distinctive theoretic framework. Gáspár (2011, pp. 93) accordingly discusses path-dependency as a frequently used concept in the social science and in general refers to the idea that events occurring at an earlier point in time will affect events occurring at a later point in time.

3. Path-dependency and Technological Development in Information Society

Strategic actions – i.e. path-making – of technological latecomers in the information society lend themselves as a core topic of sociological analysis. At the intersection of economic and sociological theory, the topic under scrutiny most notably relates to the framework of neo-institutional analysis, which focuses on classical macro-sociological issues (emergence, institutional change and preservation of institutions) by simultaneously applying a combination of economic and sociological approaches. The use of these neo-institutional analytic approaches in sociology is appropriate when explaining strategic actions in settings with incomplete information and mental models that contribute to higher transaction costs. Such conditions are regularly faced by strategic actors (Nee 1998) and it is worth mentioning that transaction costs compose a significant part of production and trade costs in contemporary economics. Thus, an important implication of alternative institutional arrangements is the fact, that they determine the direction of developmental trajectories that can cycle between economic growth, stagnation or recession (Nee 1998).

Central to the theory of neo-institutional analysis is the concept of choice within constraints. Those represent networks of interrelated norms and regulations, which define options relating to the available course of action for the actors (Nee 1998) and facilitate coordination and collective action. Norms in this sense represent a form of social capital that enable problem-solving and lead to optimal results of collective actions. In contrast, suboptimal results are the outcomes of behaviour, when individual actors do not align with the collective system as a whole and pursue the attainment of their specific goals rather than observing what is rational from the global system's perspective (individual goal rationality versus systemic rationality). Ultimately, norms that define cooperation accommodate systemic rationality and foster systemic discourse. The pivotal issue relating to developmental trajectories is whether strategic choices influence them or whether development technological development is path-dependent. Path-dependency in this context applies to processes where preceding actions determine the course of future developments. Event sequence has the ability to influence successive events in such a way that the previous path of developmental trajectory defines a narrower set of feasible future directions (Kay 2003, pp. 2). By following a "choice within constraints" approach, neo-institutional analysis in a subsidiary manner includes also path-dependency, which is common to contemporary neo-institutional analysis irrespective of a particular discipline's research domain.

When observing the previously stated definition of path-dependency, this requires specific methodological considerations of such an approach. There is no conventional body of theory or models that could be used to examine causality between variables and subsequently disprove or prove any pertaining hypotheses as is often the case in diagnostic and prescriptive research (Ostrom 1999, pp. 39). As such, path-dependency is an empirical category in its own right, used for explaining time-based processes – how they unfold and where they could potentially lead – and is increasingly gaining attention in social research community (Kay 2003a, pp. 406–407). There are no generalized explanations about developmental causality of observed systems offered by this approach. When adopting the path-dependency approach, researchers rather have to develop auxiliary "explanatory frameworks, theories and models, explaining microfoundations of path-dependent processes" (Kay 2003a, pp. 406–407), together with the technological backwardness.

It seems that the information society and the adaptation of societies to emerging information economies are particularly relevant fields of exploration (Webster 2014, pp. 19–24). This is first, due to immense dynamics that information technologies have brought to all aspects of everyday lives and secondly, due to permeability of all spheres of economy to the progress and innovations brought about by the information technology.

4. The Examples of Developmental Paths of Successful Latecomers in the Information Society – Ireland and Finland

We look at the development of Ireland and Finland as examples of successful technological latecomers that took advantage of the emerging information economy – although this aspect is often overlooked in research on these two cases' developmental path-making success. The relevance of this research is no lesser despite the problems that these countries have experienced in the recent period (financial crisis in Ireland and decline of Nokia in Finland). Their path-dependent development is analysed in three phases: diagnosis of their available choices and constraints; the critical strategic decision representing a strategic shift; and systemic discourse, which changed developmental trajectories of the two countries.

In the last 60 years, including the era of transition to the information society, only Ireland and Finland have achieved the technological development status of the most developed EU member states; in Asia, this applies to Hong Kong, Singapore, South Korea and Taiwan (also called Four Asian Tigers), whose dramatic development started in the industrial era and which have seen their highest growth between 1960s and 1990s.

Ireland and Finland started out on a level somewhere above Southern European countries and ended up joining the group of most developed European countries at the start of 21st century. Although Finland had already been relatively technologically developed and was not perceived as an European lagger like Ireland was in the beginning of the 1980s, its development was still perceived as rather dramatic, due to the ability of its national system of innovation to produce the global leader in the early era of the mobile communications, the epitome of the information society and economy.

4.1 Ireland

After 1922, when Ireland became politically independent from United Kingdom (UK), it still relied heavily on Britain. Most of Ireland's imports and exports were connected with the UK, Irish export being mostly agricultural products. Between 1922 and 1960, their main goal was to stabilize the economy, which had received an additional blow after almost half of the manufacture workforce turned instead to agriculture (Battel 2003, pp. 94–95). Because economic independence was secondary to political independence, they retained the free trade with Great Britain and did not find it necessary to open up to international markets. In fact, it was regarded by some that opening up to foreign investment and trade would represent a betrayal of national values. As a result, Ireland did not see any economic, technological or other improvements in this period.

After World War II the continental Europe saw the economic picture improve, but Ireland only received a small part of the Marshall plan resources because they did not take part in the war effort. Since the economic development policies could not be changed so dramatically and there was no way to go back to the old model, Ireland's position didn't change. It was only after the change in government in 1932, when the nationalist party Fianna Fall took over and was later replaced by Sean Lemass (Battel 2003, pp. 97–99), that Ireland started opening up to foreign markets and even formulated an Economic development programme which made Ireland an attractive foreign direct investment destination.

It was not, however, until several decades later that these measures proved results. In the meantime, Ireland went through a huge increase of public debt, inflation and unemployment rates. Nonetheless, it was too late to return to the old management model.

After the 1990s, Ireland began seeing the results of new policies and of the strategic shift, turning into an exemplary country with modernized manufacture, construction and trade. We are mainly interested in seeing how systemic discourse supported this transition by constitutionalising long-term social partnerships, which resulted not only in social partnerships but also in other, wider and more diverse effects.

4.2 Finland

Different to Ireland, Finland has never been an underdeveloped country. Finland's economy was focused on quantity instead of quality and was based on the processing of natural resources, which they had plenty of, such as pulp, paper and furniture manufacture, and on aggressive investment policy, both producing an economy with a relatively small number of companies, which were very large and vertically integrated to create an apparition of a great economy (Lilja et al. 1996, pp. 159).

Castells et al. (in Schienstock 2004, pp. 3) describe the end of the 80s and beginning of the 90s, when Finland found itself unable to weather the economic crisis – obsolete technological hardware and sudden lack of market potential due to the Soviet Union reconstruction were only some of the reasons. After focusing on technological innovations, however, Finland became a world wide leader in mobile communications and also the country with highest added value in other technology industries.

5. Coordination and Path-shaping Through Discourse

When a country wishes to reform on any level, it has to first make a "choice within constraints" and then define its path. Torfing (2001 pp. 286–287) has analysed welfare system reforms, which are known for their slow adaptation to changes, mostly because of abundant third-party interests further complicating its reformation. Even when policy-makers and stakeholders agree on changing existing policies and arrangements, they must be aware of the impact of different interests. Torfing (2001 pp. 286–287) uses the term "policy path" – not in the sense of prescribing new arrangements, but in the sense of discussion between regulatory and institutional parties for the purpose of formulating a mutually convenient policy.

In this respect, for a strategic shift to be successful, individual parties must take part in "path-shaping" to ensure a coordinated action. It should be noted, though, that old structures and arrangements can not be replaced in their entirety because of the underlying causality between them, no matter how coordinated the action of individual parties is. Nielsen et al. (1995 pp. 3–44) describe the terms "path-shaping" and "path-dependency" as complementary. While path-shaping involves changing current institutional structures and arrangements, path-dependency limits the possibilities for an exhaustive reform due to causality, as mentioned before. These two phenomena prevent an institutional vacuum and could be seen even in the rapid-changing post-socialist transition.

Offe (1995 pp. 56) mentions two problems – we must take into account when shaping a new institutional order – "hyper-rationality" and "mental residuals". The following is a solution for each of them, which should be fulfilled if we want reforms to be effective. To prevent "hyper-rationality", the designer of the new institutional order must discredit the old institutional order and show that it can no longer withstand future challenges, such as the inability of technological systems to survive a long-term economic crisis in its current form (e.g. Ireland and Finland). The second problem can be solved by alternating the perspective. Usually, reforms are not designed on the basis of their own institutional order, but rather copied from a different, successful, institutional order and adapted to the local arrangement.

For systemic discourse to be successful, the designer of institutional reforms must ensure the trust of stakeholders, who prefer a more conservative and well-defined reform. They are more apt to expect reforms to be very successful (Offe 1995, pp. 56), as defined by the designer capitalism myth (Stark 1995, pp. 67–83). No institutional reform will ever be successful, if people show no trust in it and are not prepared to support it with their own resources (Offe 1995, pp. 57).

6. Systemic Discourse in Technological Latecomers

Successful economies, such as Finland, Ireland and the four Asian tigers, show that positive development is not only achieved by redefining what will change, but also how it will be changed. Systemic discourse has thus proved to be an effective strategic process in most newly developed economies.

Different authors (O'Hearn 2000, pp. 67–92; Ó Riain 2000, pp. 157–193) write about the influence of the state as initiator, moderator and facilitator of progress, be it technological or other, whereas this does not apply only to centralized economies. Although the state is not the only partner, it must take over the initial process of assessing technological and product development to enable successful development (Castells 1997). The state must provide the framework for reallocation of resources and ensure stability of institutional order and macro-economy. That is also one of the reasons why it was the non-democratic economies that would become the frontrunners in technological development between the 60s and 80s (Castells 1997).

After the foundations of development were laid, the focus turned to more intangible resources, which required that the state connect with other partners. In Asia, the four countries began to develop systemic discourse from their centre of government towards the periphery. Systemic discourse also presented an effective developmental factor in Ireland and Finland.

6.1 Ireland

After a profound economic crisis, Irish employer's associations and trade unions began implementing social partnership agreements. The programme for national recovery (1987–1990) was the first agreement, followed by the Programme for Economic and Social Progress (1990–1993), the Programme for Competitiveness and Work (1994–1996), Partnership 2000 (1997–2000) and Programme for Prosperity and Fairness (2000–2003) (McCashin et al. 2002, pp. 266). All parties of the agreement kept to their end of the bargain and, consequently, workers were predominantly happy with the social arrangement until the general strike of nurses in 1999. Despite its weaknesses in terms of uneven distribution of social costs and increased income inequality, as presented by Kieran (2000, p. 183), it is generally accepted that the agreement benefited Ireland and contributed to its progress in the long run.

In 1993, Ireland created the National economic and social forum as a formal intermediate. At first, the forum supported communication between social partners, the state, political representatives and educational institutions and after

1997 it also allowed non-governmental organisations to join. After establishing communication between partners on a national level, they formalized regional partnerships, which facilitated networking on a more narrow level of 38 regions (McCashin 2002 pp. 267). The forum as mechanism for solving fiscal crisis was to be a mid-term facility, but turned out to be a long-term partner and factor of economic development (Hardiman 2002, pp. 17). It still exists today in the form of an institution with numerous partners involved in systemic discourse.

6.2 Finland

In Ireland technological progress was facilitated through systemic discourse, initiated with social partnership arrangements. Finland, on the other hand, tried to stimulate development by redefining the higher-education sector and supporting innovations in the field of communication, focusing mainly on local resources and ideas.

In the second half of the 20th century Finland started the reorganization process of its higher-education system by increasing its budget and workforce and establishing new institutions. Today, some of these institutions represent the driving force behind development in Finland.

In terms of innovation, Finland tried several different approaches. One was to create a network of large information communication (IC) companies with Finnish telecommunications giant Nokia at the centre. Secondly, they provided the knowledge base in the form of new technological universities and higher-level institutions. Another important step was establishing Council for scientific and technological policy, the role of which was to supply human resources, provide a nation-wide framework for innovations and stimulate technological investments. Finland also established a public research and development (R&D) agency named Tekes to fund prospective business-oriented projects. Although Tekes is a public agency, it enjoys lots of freedom in its decisions and can therefore focus on long-term effects of their funding resources. The last component of the technological progress in Finland was the risk-capital investment agency named Sitra. Together with Tekes they provided the research base as well as the financial resources for most technological projects. The success of technological development in Finland is the sum total of all these factors and the result of communication between all partners involved.

7. Not so Successful Case – Path-dependency in Slovenia

As opposed to Ireland and Finland, Slovenia is still facing a challenge of deciding its course of action. It is positioned on the EU periphery and would require large investments in infrastructure as well as revitalization and reconstruction of manufacture schemes to improve its position in the global market (Sočan 2001, pp. 53). Slovenia's goal is to be accepted among the most developed European countries known for their infrastructural development, which supports knowledge, technological development and the transfer of assets in terms of partnerships and connections. Slovenia has so far exploited all possibilities of development and will have to make a final decision regarding its future path.

Another characteristic of Slovenia is that it lacks proactivity when it comes to taking advantage of developmental instruments. Combination and coordination of developmental instruments could be a potential future path for the country, but this would require an organizational shift from hieararchy to heterarchy, so that future developmental strategies and policies are based on substance and consultation between partners. Consequently, actors on the micro and meso level would connect in the form of enterprise networks and clusters, whereas social subsystems would be able to establish an effective discourse and find a mutual agreement on future developmental path.

For Slovenia, there is no systematic empirical evidence on inter-enterprise connections and connections between social subsystems, because this research topic is not interesting. In turn, data from international studies shows that these connections are weak and limited when compared to better-developed and similarly developed EU member states, such as East and Central EU member states.

Slovenia is a typical small country based on different indicators. The downside of a small country is that some of their characteristics can deter development, but the modest size can also be an advantage. Many small European countries have become successful despite their size, though it is not possible without fulfilling certain conditions. Kuznets (1966) claimed that small countries have an economic disadvantage, which they must compensate with social development. Scharpf (2000, pp. 101–121) states that small country policies should be realistically derived from their developmental possibilities and support developmental mechanisms, while taking into account EU's integration and globalisation processes as actual obstacles.

It takes different – not more – policies and social arrangements for a semi-peripheral country to become a technologically advanced country than for a peripheral country to become a semi-peripheral one. Policies on stable macroeconomic environment or improvement of educational structure are obsolete in countries wishing to advance to a higher technological level. Choice of policies is carefully stated and any deviation can result in important social consequences.

8. Conclusion

In information societies there is a strong tendency for shifting from strategic processes, where political systems played a central role, towards joint and multilateral forms of operation, which involves a larger set of relevant systems and actors. Luhmann (1995) and Wilke (1993) attribute this phenomenon of complex societies to functional differentiation of processes. According to Wilke (1993) the new strategic process patterns are characterized by reflexivity (consideration of one's impact on the environment), contextual intervention (indirect actions, aimed at shaping external conditions and settings, instead of direct actions) and systemic discourse (searching for multilateral common grounds through joint initiatives and communication). These mechanisms enable path creation, however, they can successfully evolve only in the most developed societies. The case of successful latecomers demonstrates that certain forms of systemic discourse were key to triggering their technological breakthrough, with other factors being present sometimes even decades before. If systemic discourse is to be effective as a path-creation tool, it is therefore clear, that specific socio-cultural conditions have to be met.

9. References

Battel, Róisń Ní Mháille: "Ireland's 'Celtic Tiger' Economy". Science, *Technology and Human Values* 28(1), 2003, pp. 93–111.

Castells, Manuel: "The Power of Identity". The Information Age: Economy, *Society and Culture* 2, 1997, pp. 5–68.

Castells, Manuel / Himanen, Pekka: "The Finish Model of the Information Society". In: Schienstock, G. (eds.): *Embracing the Knowledge Economy: The Dynamic Transformation of the Finnish Innovation System*. New Horizons in the Economics of Innovation, Edward Elgar, 2004, pp. 1–23.

Domanski, Boleslaw: "Economic Trajectory, Path Dependency and Strategic Intervention in an old Industrial Region: The Case of Upper Silesia". In: Domanski, R. (eds.): *Recent Advances in Urban and Regional Studies*. Polish Academy of Sciences, Comitte for Space Economy and Regional Planning, Warsaw 2003, pp. 133–153.

Gáspár, Tamas: "Path Dependency and Path Creation in a Strategic Perspective". *Journal of Futures Studies* 15(4), 2011, pp. 93–108.

Greener, Ian: "Theorising Path Dependence: How Does History Come To Matter in Organisations, and What Can We Do About It?". University of York, Department of Management Studies, *Working Paper 3,* 2004, pp. 1–30.

Hardiman, Niamh: "From Conflict to Coordination: Economic Governance and Political Innovation in Ireland". *West European Politics* 25(4), 2002, pp. 1–24.

Jessop, Bob: "State Power". Cambridge, Polity Press, 2008.

Kay, Adrian: "Path Dependency and System Memory". Policy & Politics Conference, Policy and Politics in a Globalising World in Bristol, England, 2003, pp. 2–8.

Kay, Adrian: "Path Dependency and the CAP". *Journal of European Public Policy* 10(3), 2003a, pp. 405–420.

Kieran, Allen: "The Celtic Tiger: The Myth of Social Partnership in Ireland". Manchester: Manchester University Press, 2000, pp. 179–183.

Kuznets, Simon: "Modern Economic Growth: Rate, Structure, and Spread". Bombay: Asia Publishing House, 1966.

Liebowitz, Stan J. / Margolis, Stephen E.: "Path Dependence, Lock-In and History". *Journal of Law, Economics and Organization* 11(1), 1995, pp. 205–226.

Lilja, Kari / Tainio, Risto: "The Nature of the Typical Finish Firm". In: Whitley, R. / Hull, Kristensen P. (eds.): *The Changing European Firm: Limits to Convergence*. Routledge, 1996, pp. 159–191.

Luhmann, Niklas: "Social Systems". Stanford: Stanford University Press, 1995.

McCashin, Anthony / O'Sullivan, Eoiun / Brennan, Catherine: "The National Economic and Social Forum, Social Partnership and Policy Formulation in the Republic of Ireland". *Policy and Politics* 30(2), 2002, pp. 263–279.

Meyer, Uli / Schubert, Cornelius: "Integrating Path Dependency and Path Creation in a General Understanding of Path Constitution: The Role of Agency and Institutions in the Stabilisation of Technological Innovations". *Science, Technology & Innovation Studies* 3(2007), pp. 23–44.

Morrey, Nicola / Pasquire, Christine / Dainty Andrew / Thompson, Derek: "Path Dependency to Path Creation: Enabling Strategic Lean Implementation". Proceedings of the 20th Annual Conference of the International Group for Lean Construction 2012, pp. 1–10.

Nee, Victor: "Sources of New Institutionalism". In: Brinton, M. C. / Nee, V. (eds.): *The New Institutionalism in Sociology*. Stanford: Stanford University Press, 1998, pp. 1–16.

Nielsen, Klaus, Bob Jessop and Jerzy Hausner: "Institutional Change in Post-Socialism". In: Hausner, J. / Jessop, B. / Nielsen, K. (eds.): *Strategic Choice and*

Path-Dependency in Post-Socialism: Institutional Dynamics in the Transformation Process. Aldershot: Edward Elgar, 1995, pp. 3–44.

Offe, Claus: "Institutional Design, Strategic Dilemmas, and the Dynamics of Post-Socialist Transformation". In: Hausner, J. / Jessop, B. / Nielsen, K. (eds.): *Strategic Choice and Path-Dependency in Post-Socialism: Institutional Dynamics in the Transformation Process.* Aldershot: Edward Elgar, 1995, pp. 47–66.

O'Hearn, Dennis: "Inside the Celtic Tiger: The Irish Economy and the Asian Model". Dublin: Pluto, 1998, pp. 67–92.

Ó Riain, Seán: "The Flexible Developmental State: Globalization, Information Technology, and the 'Celtic Tiger'". *Politics & Society* 28(2), 2000, pp. 157–193.

Ostrom, Elinor: "Institutional Rational Choice: An Assessment of the Institutional Analysis and Development Framework". In: Sabatier, P. (eds.): *Theories of the Policy Process.* Boulder, CO. Westview Press, 1999, pp. 35–71.

Scharpf, Fritz: "Interdependence and Democratic Legitimation". In: Pharr, S. J. / Putnam, J. R. (eds.): *Disaffected Democracies: What's Troubling the Trilateral Countries.* Princeton, New Jersey: Princeton University Press, 2000, pp. 101–121.

Sočan, Lojze: "Nova ekonomija in Slovenija". [New Economy and Slovenia] In: Stanojević, M. (eds.): *Uspešna nedozorelost. [Successful Immaturity]* Ljubljana: Fakulteta za družbene vede, 2001 pp. 49–72.

Stark, David: "Not by Design: The Myth of Designer Capitalism in Eastern Europe". In: Hausner, J. / Jessop, B. / Nielsen, K. (eds.): *Strategic Choice and Path-Dependency in Post-Socialism: Institutional Dynamics in the Transformation Process.* Aldershot: Edward Elgar, 1995, str. 67–83.

Sydow, Jörg / Schreyögg, Georg / Koch, Jochen: "Organizational Paths: Path Dependency and Beyond". 21st EGOS Colloquium, Berlin, Germany 2005, pp. 2–3.

Torfing, Jacob: "Path-Dependent Danish Welfare Reforms: The Contribution of the New Institutionalisms to Understanding Evolutionary Change". *Scandinavian Political Studies* 24(4), 2001, pp. 277–309.

Webster, Frank: "Theories of the Information Society". New York: Routhledge, 2014.

Wilke, Helmut: "Systemtheorie". Stuttgart: Gustav Fischer Verlag, 1993.

Dolores Modic
Faculty of Information Studies in Novo mesto
Ljubljanska cesta 31a, p. p. 603, 8000 Novo mesto, Slovenia
dolores.modic@fis.unm.si

Intellectual Property Rights in the Information Sector: Sociological Perspective

Abstract: *In this paper, we make a case for adopting a sociological approach to studying intellectual property rights in the information sector. The field has been in the past strongly dominated by researchers of other backgrounds and sociologists have for a long time not offered much to the study of intellectual property rights. We intend to remedy this. We offer insight on how we can utilize the sociological approach, sociological theories and social analysis by first focusing on the post-positivist approach and its utility for the intellectual property studies in the information sector. Secondly, we look into the usefulness of the innovation systems theory, also pointing out the problems one might have when using it. Thirdly, we bring at the forefront the fuzzy-set (fsQCA) as a method of a Qualitative Comparative Analysis (QCA) and contemplate on its benefits as well as warn researchers to use it with due diligence. By delving into each of the abovementioned topics, we show what sociology has to offer to studies of intellectual property rights inside the information sector.*

Keywords: *sociological analysis, sociological theory, intellectual property rights, mixed methods, fuzzy-set*

1. Introduction: Information Industries and Intellectual Property Rights Therein

There are many discussions related to what some have coined together as the Age of information and communication technologies. As Freeman and Louca (2002, p. 301) have written "even those who have disputed the revolutionary character of earlier waves of technological change, have little difficulty accepting that a vast technological revolution is now taking place, based on computer software, microelectronics, the internet and mobile phones". Webster (2014, pp. 10–11) has written that there are several possible criteria to define how the "new" information society is different from before: technological, economic, occupational, spatial and cultural. Looking at technological (with its emphasis on the information communication technologies (ICT)) and economic (focusing on information sectors and their contribution) criteria we can note significant problems of taxonometry and disambiguation.

From the time of the first pioneers in this field - such as Porat and Machlup – more or less encompassing categories of information industries have been created – some of them also exaggerating their economic worth. *Inter alia* OECD has, in the last decade, focused on the disambiguation challenges of the ICT sector defining it as encompassing ICT manufacturing industries, ICT repair industries, ICT trade industries and ICT services industries (OECD, 2007, p. 12). It also looked into the definition of the information sector (ibid.), which encompasses the so-called content sector - or in other words the creative sector if we adopt the North American Industry Classification System (NAICS) terminology.

Our definition of the information sector is a broad one, derived from a technical point of view (and aligned with prevailing definitions). We adopt the definition that the sector comprises of actors engaged in the following processes: (a) producing and distributing information and cultural products (hence, also encompassing creative industries); (b) providing the means to transmit or distribute these products as well as data or communications; and (c) processing data (BLS 2014; UNSTATS, 2015).

Let us point out some initial facts on intellectual property rights (IPRs) in the information sector (and beyond). Firstly, mechanisms for intellectual property (IP) protection may be distinguished to formal (patents, models, designs, trademarks, etc.) and informal mechanisms (trade secrets, first on the market, complementary goods, etc.). Secondly, a lot has been written in recent years on the justification of IPR (and especially patents) – where the traditional notion of IPR's as drivers of innovation is often disputed and IPR's are seen as tools for appropriation. But companies encounter the problem of the low effectiveness of all (formal and informal) mechanisms (Cohen et al. 2000; Levin et al. 1987; Neuhaeussler 2009). IPRs are riddled by uncertainties and we often speak of probabilistic IPRs (Modic / Nadij 2016). The accompanying issue is hence also that of the appropriability of intellectual property in the sense of possibilities of protecting the intellectual property from (unlicensed) imitation and of reaping profits from these.

Nonetheless, intellectual property rights have seen a surge in recent decades. This is mirrored in the rising numbers of applied and granted intellectual property rights (this simple fact may be seen from practically any IPR database; World Intellectual Property Organization's (WIPO's), European Patent Office's (EPO's), Office for Harmonization in the Internal Market's (OHIM's), etc.). However, it seems that a relevant question of why that is, remains unanswered in its fullness (though please note several attempts to explain this issue). From the Organisation for Economic Co-operation and Development (OECD) Patent Database we can see a rise in patents and patent applications inside the ICT sector as well, which can be seen for practically all relevant categories (Patent Cooperation Treaty (PCT) applications, US patent grants etc.) between 1999–2011/12.

Information industries comprise of actors primarily engaged in producing and distributing (except by wholesale and retail methods) information and cultural products. The intangible nature of the content of information and cultural products allows for their more uncontrollable diffusion inside the system, hence the topic of intellectual property rights is an especially relevant one. The formal (legal) mechanism preferred inside information industries is the copyright, which is usually not seen as providing a solid protection against infringement. The other formal mechanisms of choice are patents.

We also need to look to the issue of so-called software (or computer) patents. Some jurisdictions allow such patents to be applied for and granted, whereas others do not allow it – at least not officially. But note that although e.g. the European Patent Office officially does not allow for them to be granted, there were already more than 30.000 European software patents in existence by 2002! These present a special feature of IPRs inside the information sector.

Companies inside (and beyond) the information sector often do not possess sufficient knowledge to use IPRs as efficiently as possible. It is hence not surprising that e.g. Tang (2007) discovered for electronic publishers that they not only do not patent, but that the majority also does not know that software-based applications could be patented. Furthermore, intellectual property rights are labeled as ineffective, partly due to their sub-optimal use by the companies (see also Modic / Nadij 2016).

Another feature of the new information age/society and its implication for the information sector and the intellectual property rights therein needs to be pointed out. Some authors claim that: "the computer age is over" (Gilder 2000, p. 1) or in other words that Gilder's "microcosmus" is ending and in its place the age of "telecosmus" is born. Although the influence of computers and its components remains strong, this new environment has led to organizational changes as well. It is a rather complicated environment for intellectual property rights and their utilization, which are now coming to the forefront. In the world of today, companies do not innovate anymore in isolation from other actors (Edquist 1997 [2005]), but are entering into (innovation) relationships with other actors. Furthermore, actors' ideas are quickly combined with other ideas and/or re-used. The outcome of this kind of environment is characterized by a large percentage of adaptations; a situation welcomed by authors of innovation systems policies, but frowned upon by individual companies due to their appropriative aspirations dominating the use of IPRs. Issues, especially important to the ICT (as well as other complex industries) sector (as a sector relying upon products that are characterized by their intangible nature) are hence also those of patent portfolios, patent thickness, patent trolls, etc. All of these affect the relationships between actors in the system.

This is the *status quo* in which we need to employ our research efforts. An environment in which a comprehensive and holistic approach is very much appreciated. We continue by making a case for a post-positive approach, the use of innovation studies theories (or framework) and the fuzzy-set method, thereby illustrating the potential value of sociology for studying intellectual property rights in the information sector.

2. Briefly on the Post- Positive Approach and Intellectual Property Rights

Inside intellectual property rights studies the knowledge base – lying in the very heart of intellectual property – is composed not only of knowledge embodied in different intellectual property rights, but also in tacit knowledge, know-how, practical experiences, research materials, etc. This is a strong argument in favor of what Foray and Mairesse (2002, p. 43) call the uncontrollability in this field and high levels of uncertainty. However, a sociological approach and sociological analysis are offering us tools, so we may try to fight ourselves through and (at least to a point) conquer this uncontrollability.

Swedberg (2014, p. 8) has written (following Peirce 1929) that science is about observing a phenomenon, coming up with the idea or a theory why something happens, and then testing the theory against facts. Both authors also claim that it is the abducation, or coming up with an idea, which is the most important part of the whole research process. It is also where we may find comparability with innovation studies – both approaches acknowledging the vital importance of the ideas itself. However, we must note, that the idea is of little value if not tested against facts as is an innovation (or better said invention) of little value if it cannot survive the test of surviving on the relevant market.

Intellectual property (rights) studies have for a long time relied on positivist approaches, which have had its limitation especially when looking at intellectual property rights inside today's complex innovation environment. However, post-positivism – as an attempt to transcend and upgrade positivism (Adam 2014) – brings significant steps towards understanding how and why intellectual property rights work (or do not work) inside the information age and the networked society of today. We see this as the new direction in the studies of intellectual property rights, where mere interpretation of (larger) quantitative dataset is now transcended.

Sometimes we may see as a prevailing idea that the methodology of post-positivism is the qualitative methodology on the account of rejecting quantitative methodology. We believe this not to be true and that post-positivism simply tries to harness it within a more complex research design. One could describe it as more cautious concerning strong and one-sided interpretations and restrained regarding the too extensive (or obsessive) use of (quantitative) data and methods (see also Adam 2014; Achen 2005).

It would seem this postpositive approach is very useful for innovation studies. Innovation itself has been seen as »the emblem of the modern society and a panacea for resolving many problems« (Godin 2015, p. 1) or in other words it seems the »innovation mantra« is ever-present. The term innovation usually depicts disruptive, unique and unprecedented phenomena, enhancing the need for strong, well suited and well thought out theoretical frameworks encompassing innovation studies. Intellectual property rights seem to be under researched in terms of more holistic views; which are just starting to pop-up in recent years. Statistical, econometric (especially in economy) and case study (especially in legal sciences) approaches have been in the forefront of intellectual property rights research, with patent, trademark or design database analysis and analyzing datasets gathered by (large or small sample) surveys. Now the time is ripe for new approaches as well as delving in in-depth issues of patenting and the use of other IPRs. Post-positivism with its view that knowledge is therefore not absolutely certain, but conjectural and the understanding of the world is open to the possibility of future refutation, seems to fit especially well with the IPR field, which is riddled with uncertainties, uncontrollability and lack of knowledge of the underlying mechanisms.

The post-positivist approach hence reminds us of underlying processes and guides us to diligently study all features that can affect the IPR results as well as to take special care of uncovering reasons that lie behind anomalies - since these may lead us to discovering new facets and features previously not taken into consideration. The IPR field is known for its uncontrollability and high level of uncertainties, thus it is of utmost importance we try to gain a holistic view on the IPR processes and relationships.

In the next part we highlight the usefulness of innovation systems theories, which we see as a possible appropriate framework for intellectual property studies, where it becomes somewhat clear we are better off seeing the (scientific) theory as an evolving institution (Barnes et al. 1996, p. x).

3. Usefulness of Innovation Systems Theories for IPR Studies

Scott (2011, p. 1) claims that "a diversity of theories and perspectives is widely seen as an essential characteristic of sociology and, perhaps, an especially marked feature of contemporary sociology". Also approaches to innovation studies are based on the intersection of social studies of science and technology and economic sociology, as well as theories of national innovation or learning systems. This fact brings a more interdisciplinary approach to innovation studies (as well as possibly to IPR studies); applying economical, legal or other methods and research tools.

The boom of works in the field of innovation systems has been immense in the last decade. One must keep in mind that a system for production of technological innovations is primarily a social system. Learning and knowledge are central to innovation processes and involve interactions between different actors in markets as fields where actors meet (Beckert 2010). A general definition of innovation systems focuses on "elements and relations which interact in the production, diffusion and use of new, and economically useful, knowledge" (Lundvall 1992). However, the term "systems of innovation" is a "genotype" term with a number of "phenotypes" at different levels: national, regional or local systems of innovation, as well as sectoral, technological and metropolitan systems of innovation (Rončević / Modic 2012). Their joint premises is, that different factors exist, which influence the development, diffusion and exploitation of innovations and that the key to enhancing innovation performance are the relationships between different actors of the innovation system. They are based on the assumption that location and spatial proximity matter for innovation activities (Porter, 1990; Feldman and Massard, 2002). In other words innovation systems are comprised of three structural elements: actors, institutions and transfer of knowledge and technology (cooperation).

We first highlight two general problems that seem to hinder the innovation systems theories' (or as some note of innovation systems "conceptual frameworks" – e.g. Edquist / Hommen, 2008) better usefulness for intellectual property right studies.

The first issue we need to look into is the one of the concept itself. Today we have a plethora of partially overlapping concepts at our disposal (e.g. D'Allura et al 2012, p. 141–142; Carayannis 2014, p. vi-vii) and at the same time a lack of a coherent theoretical, conceptual and methodological approach to study them (Rončević 2012). However, Isaksen (2001, p. 107) sees the difference between other concepts and innovation systems in their more "planned" nature as opposed to a more "spontaneous" nature of other concepts, giving the possibility for rational action. The latter also gives us hope in terms of the usefulness of the systems of innovation theories for innovation studies. The second issue is that of the definition of terms used. In the last two decades the semantics of innovation systems theories has almost overflowed. The sociological study of innovation and innovation systems theories show us a fractured definition of both main terms: innovations and institutions. The term "innovation" is used by sociologists to refer to the creation of new ideas (inventions) as well as their dissemination and commercial success and diffusion. Additionally, the term can refer to new organizational forms or new organizational processes and routines. The terminology makes neither distinctions between innovations and inventions nor many times between innovations and adaptations/imitations. The second term – usually causing quite a stir is institution. Some authors define institutions in terms of institutional rules, some define them as actors. One is obvious; these nebulous and numerous definitions are hardly helpful for analyzing IPRs. The problem being, if phenomena whose changes are to be explained are qualitatively inhomogenous, the *de facto* value of such theories using these terms plummets. Hence, the researcher must start its research when using the innovation systems theories (framework) with rigorous and thorough definition of the used terms.

The above mentioned notwithstanding, we can derive considerable benefits when using the innovation systems framework for studying intellectual property rights (we rely also on Sharif's advantages (2004, p. 86) for using innovation systems framework for innovation studies).

First, the system of innovation framework goes beyond research and development (R&D) in an effort to explain innovation dynamics. One obstacle in understanding what innovations and IPRs are, is derived from the fact that they have been studied by different communities of researchers with different backgrounds and the failure of these communities to effectively communicate with each other. Since the term intellectual property rights mean different things to researchers of different backgrounds, they focus on different facets of intellectual property and on different intellectual property mechanism. When legal researchers study intellectual property, they will focus mostly on innovations that are new, and the issues of adaptations will usually be taken into account only when speaking of intellectual property protection violations. Furthermore, the issue of the true efficiency of intellectual property rights is usually not included in the legal scholars' research scope (at most they are concerned with legal issues connected to intellectual property protection processes in case of judicial adjudication of a violation). On the other hand, economists do tend to focus on adaptations, especially when dealing with intellectual property transfers, but are less apt in recognizing legal issues encompassing them.

The issue of appropriation and knowledge/technology spillovers comes to the forefront inside the innovation systems and with them various mechanisms that protect intellectual property and/or allow the transfer of intellectual property from one actor to the other. *Vice versa*, the innovation systems theory and/or framework give rise to the fact that intellectual property rights need to be analyzed not only on the organizational level but also different features of said organizational environment need to be taken into account. This allows them to focus on the relationships and how these effect the efficiency of intellectual property rights (IPRs) - firstly as appropriation mechanisms and secondly as mechanisms allowing or preventing the diffusion of innovations.

Secondly, innovation systems framework takes into account that learning processes are very much influenced by institutional set-ups that foster competition and co-operation. One of the problems of intellectual property utilization is the lack of knowledge connected to IPRs. Through this it puts into focus the relationships between individuals (a new budding area might be one of so-called organization innovation systems - although we can see it still in its infancy), since much of knowledge transfer regarding the knowledge on IPRs itself (and not their underlying innovations) is done not through formal channels (acquiring formal education on IPRs) but *via* apprenticeship. In this environment it is of key importance we have the framework that allows us to disentangle the processes and influences that effect the creation or acquisition of knowledge needed for the efficient use of IPRs. The innovation systems' approach is also well suited to analyses of technology policy, since it draws attention to the systematic features of the innovation process and their variation.

Finally, the innovation systems framework allows the emphasis to shift away from the organization as the sole vector of technological innovation in a society to the role of government policy, legal institutions, education and training institutions, and even norms and regimes. Innovation property rights are specified and enforced by political institutions, and the literature simply takes those as a given. Similarly can be said for the preposition intellectual property rights are effective and efficient. Looking at these issues inside the innovation system framework through the lenses of an economist, they do look into the efficiency; but usually with a specific focus (whether the unit of research is a company, a group of (similar) companies, a state, etc.) and without much regard for what is »hiding« behind these relations between actors. This is where a sociological viewpoint comes into the forefront, since looking into these relations seems to be the main distinguishing point of defining a specific worth of sociology for the analysis of IPRs. They let us acknowledge that organizations are embedded in their institutional environment; however it is the organizations that produce institutions. This is why Edquist (1997 [2005]) speaks of mutual embeddedness. In this sense the innovation systems framework allows us to evaluate on the macro level the role of the intellectual property rights inside the broader institutional set-up and on the micro level to evaluate relevant features that affect the use of IPRs.

Hence, the value of using the innovation system theory is in shifting the focus towards relationships inside and outside the studied organizations (also regions, etc.) and on the institutional framework that encompasses it. We believe that without those *foci* the value of IPR studies remains sub-optimal.

4. Sociological Analysis of IPRS within THE Information Sector: A Case for the Fuzzy-Set

The sociological analysis of social systems of innovation is still underdeveloped (Rončević and Modic 2011). Economics, economic and regional geography and political science are in the forefront of this strand of research. However, we have already written elsewhere that sociology has much to offer in the analysis of institutions (ibid.) and innovation systems in general as well as to innovation studies. The sociological approach to the study of innovation attempts to examine the way that social structure influences both the process and products of innovative activity (Mako Hill 2010).

Few social scientists dispute the need to combine qualitative and quantitative methods and evidence. The real question we need to ask ourselves is how to do so. Achen (2005, p. 30) has noted that "first-rate social science theorizing seems to integrate the two in ways we do not fully understand." Let us see, how social analysis may come in handy, when looking at issues of intellectual property rights in the information sector.

As said previously, the predominant IPRs used by the actors of innovation system in the information sector are copyright and patents. The former we can measure; the patent indicators and their values can be easily found (together with some other, many times useful, data) in patent databases. On the other hand copyright does not require registration (at least not necessary registration) - these data can be found in (previous or own) survey results (again, usually equipped with at least some additional useful data). The latter is making research on IPRs in the information sector more difficult, since we are sometimes faced with an initial lack of basic data. We also noted the need to look to the issue of so-called software (or computer) patents *versus* other ways to protect software innovation. These phenomena must be taken into account, when looking into how one will approach studying the patents and mining the patent databases inside the information sector. In any case, there can be no doubt that patent (as well as trademarks and design) databases analysis exhibit numerous advantages as data sources. They have correspondingly provided the basis for the progress of our understanding of not only the use of IPRs, but also of innovation, technological and economic progress as well as the diffusion of knowledge.

However, this kind of predominantly quantitative analysis of data deprives us of in-depth understanding of the underlying issues. What we are many times really interested, in is relationships. If for example, a company patents heavily, what we could potentially be interested in, is whether they "produce" successful or unsuccessful patents or even more interestingly, which are its organizational features (or/and the combinations thereof) or the institutional set-up that allow them to produce patents of a certain quality. Or to give another example, the database analysis may show us, the longer the firm is in existence, the more likely it is it will patent or will engage in IPR transfers. But the question, as to why this is so (is it because their information-processing capabilities or their coordination capabilities have risen, their incentive alignment is stronger, or simply because they have more financial means at their disposal), remains unanswered. Real institutions are a combination of formal rules and actual social practices. This lies in the heart of dilemmas regarding IPRs in the information sector.

Additionally most measures in social science are uncalibrated. They indicate the cases' positions in distributions relative to each other, but not relative to a known standard. Looking into the above-mentioned examples what we would need to do, is find a way to combine the quantitative data from the databases with the qualitative data acquired in the effort to try to get a grip of the in-depth issues. But qualitative research on IPRs is mostly limited by the sample size and predominantly a limited sample or case studies are utilized.

One idea – which is also our research proposal for anyone engaged in researching the above-mentioned topics as well as anyone who wishes to go beyond what quantitative data on intellectual property rights can tell us and look into in-depth issues (engaging in intensive and integrative research of how different parts of a case fit together) – is to examine if it is possible to take advantage of the so-called Qualitative Comparative Analysis (QCA).

The QCA is an analytic technique that uses Boolean algebra to implement principles of comparison. By formalizing the logic of qualitative analysis, QCA makes it possible to bring the logic and empirical intensity of qualitative approaches to studies that embrace more than a handful of cases - research situations that normally call for the use of variable-oriented, quantitative methods (Ragin 2006). The most used method is usually now – following Ragin's terminology – in sociology discussed under the title of "fuzzy-set" (fsQCA).

The method differs from traditional regression analyses, since it is based on theory and logic and not statistics, which embeds the method inside the post-positive approach. It is designed to evaluate social systems characterized by causal complexity. Consequently, fuzzy-set assumes there can be many pathways to the same outcome, a phenomenon known as equifinality. The method hence looks for the effect of combinations of necessary and sufficient explanatory characteristics, rather than for the effect of each individual characteristic. This is especially important in cases when investing in-depth problems of IPRs, where there is always a unique innovation (hence by definition something that is novel and different from the rest) in the centre. Following this, our initial instinct is, it seeks a case based approach. But case-based approaches seem to be lacking due to their weaker comparability value, which is where methods, such as the fuzzy-set, come into play.

As to the broader question why the mixed method approach (or triangulation – a term often used inter-changeably) has not received sufficient attention in the past, lies in the perception that qualitative and quantitative research are distinct, since they are said to be based on different philosophical principles. Johnson and Onwuegbuzie (2004) claim that pragmatism is the "philosophical partner for mixed methods research".

Niglas (2009) however warns that there seems to be "an increasing pressure and an aroused but "immature" readiness for postgraduate students and novice researchers in many fields to use a mixed methods approach for their research projects without further questioning if their work really benefits from it". One of the ways a novice researcher could see if a fuzzy-set method in the field of intellectual property research is appropriate would be by following the Onwuegbuzie and Sutton (2006 in Onwuegbuzie and Leech, 2009) 13-step concept to mixed research process. For example such a study could be done on the above example of researching the successfulness of patenting in the information sector by looking into companies in the information industries that patent heavily combined with the issues of whether they "produce" successful or unsuccessful patents and, which are the organizational features that cause them to produce patents of a certain quality. The concept allows for a step-by-step validation of the usefulness of the mixed method approach (e.g. determining the mixed goal of the study and the mixed research objectives; combining quantitative analysis of patent data in the information sector to gain as relevant qualitative sample as possible with a subsequent qualitative analysis etc.).

Let us re-iterate: it is our opinion that mixed methods increase the credibility and validity of the results and we see the fuzzy set (fsQCA) as an especially valuable method. In general an important contribution of sociological analysis is that it allows for sufficient flexibility and provides appropriate analytical (concepts and) tools to enhance our knowledge of IPRs as used inside the information sector.

5. Conclusion

Lester and Piore (2006, p. 5) believe that we are in danger of learning the wrong lessons about innovation. We acknowledge this to be true, if we are looking into intellectual property rights and their use inside the information sector. Fortunately, the post-positivist mantra, slowly permeating the social sciences (and especially sociology), has given us a new view on how to handle IPR data and what we may learn from it. The post-positive approach is coming to the forefront also due to the - now already established - ideas there is a variety of different paths and combination of features that can lead to (un)successful IPRs. It is allowing for sufficient flexibility by asserting that knowledge does not come only from past observation and evidence, but also from inspiration and innovation. It allows the researcher to delve in depth into IPR topics that are riddled by probabilities and where there is still much to do to diminish the uncontrollability and uncertainties inside the IPR field.

Looking at the use of mixed methods inside the research of intellectual property rights in the information sector (or beyond), we have until now not seen an abundance of them. There are still many un-answered questions on how to truly utilize methods and tools of social analysis to enhance our knowledge on the use of IPRs inside the information sector. Due to this fact we have shown not only the benefits of using these methods, but also gave guidelines on when to use them. We have done so in order to contribute to what Etzioni (2007, p. 157) sees as giving a much needed "focus on the instruments to be utilized to elevate the analysis of societal issues /and/ to improve on amateur, intuitive, or journalistic sociology." It is our belief that sociology can truly enhance the quality of IPR studies *via* rigorous and thought out application of its tools and methods.

We have made a case for the use of fuzzy-set (fsQCA). We consider it a very useful method in research of IPRs, since it helps diminish criticisms that generalization of qualitative results is unfeasible (while individual case studies are very present in IPR research). It also allows overcoming weaknesses of quantitative setups (relying on pure IPRs' database analysis), since an in-depth focus into relevant inter-related issues is made possible, which allows for a more comprehensive understanding. The method is especially valuable inside the information sector with its special features (especially heavy reliance on copyright and issues related to software patents), causing a lack of available IP database data. The fuzzy-set also relies on equifinality; aligning it with the above-mentioned idea that there are many paths to (un)successful use of IPRs.

We also highlight the usefulness of innovation systems theories for the research of IPRs in the information sector (and beyond). Its value is in directing attention to underlying processes, relationships as well as to the framework conditions. This is especially important in the studies of IPRs since the true value of IPRs does not stem solely from the company that holds the IPR itself, but is dependent on reactions (actions) of other actors inside the innovation system to the IPR and their restrictive (negative) nature. The framework conditions greatly influence the use and efficiency of IPRs, which is also one of the features the use of innovation systems theories can point us to. Furthermore, we not only show the benefits of using innovation systems theories, but also point out some restrictions, so the researchers may utilize them more efficiently. This paper gives us an initial glimpse on some dilemmas inherent to research in this area. Hopefully, other authors shall offer even more insight combined with useful suggestions.

6. References

Achen, Christopher H.: "Two Cheers for Charles Ragin". *Studies in Comparative International Development* 40(1), 2005, pp. 27–32.

Adam, Frane: *Measuring National Innovation Performance: The Innovation Union Scoreboard Revisited*. Springer: Heidelberg et al., 2014.

Barnes, Barry / Bloor David / Henry, John: *Scientific Knowledge: A Social Analysis*. Chicago University Press: Chicago, 1996.

Beckert, Jens: "How do Fields Change? The Interrelations of Institutions, Networks, and Cognition in the Dynamics of Markets". *Organization Studies* 3(5), 2010, pp. 605–27.

BLS (Bureau of Labor Statistics): *Industries at a Glance: Information: NAICS 51(2014)*, retrieved 4.1.2015 from: http://www.bls.gov/iag/tgs/iag51.html.

Carayannis, Elias G.: "Series Foreword". In: Betz, Frederick (ed.): *Societal Dynamics. Understanding Social Knowledge and Wisdom*. Springer: Heidelberg et al., 2014.

Castells, Manuel: *The Rise of the Network Society: second edition*. Wiley- Blackwell: Oxford, 2010.

Cohen, Wesley M. / Nelson, Richard R. / Walsh, John P.: *Protecting Their Intellectual Assets: Appropriability Conditions and Why U.S. Manufacturing Firms Patent (or Not). (NBER working paper 7552).* National Bureau of Economic Research: Cambridge, 2000.

D'Allura Giorgia, M. / Galvagno, Marco / Mocciaro Li Destri, Arabella: Regional Innovation Systems: A Literature Review. *Business Systems Review* 1(1), 2012, pp. 139–156.

Edquist, Charles / Hommen, Leif: "Comparing national systems of innovation in Asia and Europe: theory and comparative framework". In: Edquist, Charles / Hommen, Leif (eds.): *Small Country Innovation Systems: Globalization, Change and Policy in Asia and Europe.* Edward Elgar: Cheltham et al., 2008, pp. 1–30.

Edquist, Charles: "Introduction". In: Charles Edquist (ed.): *Systems of Innovation: Technologies, Institutions and Organizations.* Routledge: London et al., 1997[2005], pp. 1–35.

Etzioni, Amitai: "Social Analysis and Social Action". In: Dasgupta, Samir / Driskell, Robyn (eds.): *Discourse on Applied Sociology: Theoretical Perspectives.* Anthem Press: London et al., 2007, pp. 157–164.

Feldman, Maryann / Massard, Nadine: "Location, Location, Location: Institutions and Systems in the Geography of Innovation". In: Feldman, Maryann / Massard, Nadine (eds.): *Institutions and Systems in the Geography of Innovation.* Kluwer Academic Publisher: Massachusetts, 2002, pp. 1–20.

Foray, Dominique / Mairasse Jacques.: "The Knowledge Dilemma and the Geography of Innovation". In: Feldman, Maryann / Massard, Nadine (eds.): *Institutions and Systems in the Geography of Innovation.* Kluwer Academic Publisher: Massachusetts, 2002, pp. 35–54.

Freeman, Chris / Louca Francisco: *As Time Goes By: From the Industrial Revolution to the Information Revolution.* Oxford University Press: Oxford, 2002.

Gilder, George: *Telecosm: How infinite bandwith will revolutionize our world.* Touchstone: New York, 2000.

Godin, Benoit: *Innovation Contested: The Idea of Innovation over the Centuries.* (New Routledge Studies in Social and Political Thought). Routledge: New York, 2015.

Isaksen, Arne: "Building Regional Innovation Systems: Is Endogenous Industrial Development Possible in the Global Economy". *Canadian Journal of Regional Science/Revue canadienne des sciences régionales* XXIV(1), 2001, pp. 101–20.

Johnson, R. Bruce / Onwuegbuzie, Anthony J.: "Mixed Methods Research: A Research Paradigm Whose Time Has Come". *Educational Researcher* 33(7), 2004, pp. 14–26.

Lester, Richard K. / Piore, Michael Y.: *Innovation: The Missing Dimension.* Harvard University Press: Harvard, 2006.

Levin, Richard C. / Klevorick, Alvin K. / Nelson, Richard R. / Winter, Sidney G.: "Appropriating the Returns from Industrial Research and Development". *Brookings Papers on Economic Activity*, 1987(3), 1987, pp. 783–831.

Lundvall, Bengt Ake (ed): *National Systems of Innovation: Towards a Theory of Innovation and Interactive Learning*. Pinter Publishers: London, 1992 [2010].

Mako Hill, Benjamin: *The Sociology of Innovation. General Examination Administered by Jason Davis* 2010, retrieved 4.1. 2015, from: http://mako.cc/academic/generals/bmh-generals-sociology_innovation.pdf.

Modic, Dolores / Damij, Nadja: "Own-it": Managing Intellectual Property Processes via the Activity Table in Creative Industries." In: Lugmayr, Artur (ed.): *Information Systems and Management in eMedia Industry*. Springer: Berlin, 2016-fothcomming.

Neuheusler, Peter: "Formal vs. informal protection instruments and the strategic use of patents in an Expected-Utility framework". (Fraunhofer ISI Discussion Papers Innovation Systems and Policy Analysis, No. 20). Fraunhofer ISI: Karlsruhe, 2009.

Niglas, Katrin: "How the novice researcher can make sense of mixed method design?". *International Journal of Multiple Research Approaches* 2009(3), 2009, pp. 34–46.

OECD: *Information Economy Sector Definitions Based on the International Standard Industry Classification (ISIC 4)*. (Paper DSTI/ICCP/IIS(2006)2/FINAL). OECD: Geneva, 2007.

Onwuegbuzie, Anthony J. / Leech, Nancy L.: "Lessons learned for teaching research methods research: A framework for novice researchers". *International Journal of Multiple Research Approaches, Vol.* 2009(3), 2009, pp. 105–107.

Porter Michael E.: *The Competitive Advantage of Nations*. The Free Press: New York, 1990.

Ragin, Charles C.: *User's Guide to Fuzzy-Set/Qualitative Comparative Analysis 2.0*. Department of Sociology, University of Arizona: Tucson, 2006.

Rončević, Borut: "Regional development agencies and changing social fields; towards a sociology of regional systems of innovation". In: Bellini N, Danson

M / Halkier H (eds.): *Regional development agencies: the next generation?: networking, knowledge and regional policies*. Routledge: London et al., 2012, pp. 87–101.

Rončević, Borut / Modic, Dolores: Regional systems of innovations as social fields. *Sociologija i prostor* 49(191), 2011, pp. 313–333.

Rončević, Borut / Modic, Dolores: "Social fields of technological innovations". In: Genov N (ed.): *Global trends and regional development, (Routledge studies in development and society, 28)*. Routledge: New York, 2012, pp. 226–247.

Scott, John: *Conceptualizing the Social World: Principles of Sociological Analysis*. Cambridge University Press: Cambridge, 2011.

Sharif, Naubahar: "Contributions from the Sociology of Technology to the Study of Innovation Systems". *Knowledge, Technology, & Policy* 17 (3–4), 2004, pp. 83–105.

Swedberg, Richard: *The Art of Social Theory*. Princeton University Press: Princeton, 2014.

Tang, Puay: "Management of intellectual property by electronic publishers". In: Blackburne, Robert A. (ed.): *Intellectual Property and Innovation Management in Small Firms*. Routledge: London et al., 2007, pp. 92–113.

United Nations Statistics Division (UNSTATS): *Information Sector*, retrieved 04.01.2015 from:http://unstats.un.org/unsd/cr/registry/docs/i31_information.pdf.

Webster, Frank: *Theories of the Information Society: Fourth Edition*. Routledge: London et al., 2014, pp. 1–20.

Erika Džajić Uršič and Borut Rončević
School of Advanced Social Studies in Nova Gorica
Gregorčičeva 19, 5000 Nova Gorica, Slovenia
erika.ursic, borut.roncevic@fuds.si

Industrial Symbiotic Networks in the Information Society: Research Challenges and Perspectives

Abstract: *The authors conceptualize industrial symbiotic networks and propose an interdisciplinary research program. Industrial symbiotic networks, whose prominence overlaps with the speeding-up of information society, transcend the boundaries of various systems, which includes the social system. This implies that they are not only technological networks, but are also social networks. The suggested research program falls within the realm of computational sociology and should model the morphogenesis of a complex network of networks: technological, informational, logistical and social.*

Keywords: *industrial symbiosis, networks, research strategy, simulations, computational sociology*

1. Introduction

This paper will conceptualize industrial symbiotic networks (ISNs) and outline the proposal of an interdisciplinary research program. Industrial symbiotic networks are dynamic, evolving, complex networks that are transcending the boundaries of technical, natural computer and social systems. These came into the focus of scientific research with research on industrial symbiosis since the beginning of the 1990s.

Industrial symbiosis is a paradigm of connecting flows of services, resources, by-products and waste. It emerged as a set of exchange structures to advance to a more eco-efficient industrial system, by establishing inter-organizational networks of waste material and energy exchanges. The research on industrial symbiosis gained prominence with increasing interest in industrial ecology. The relationship between industrial symbiosis and industrial ecology is not completely straightforward and has been conceptualized in a variety of ways. According to some conceptualizations, industrial symbiosis is one of the most successful forms of industrial ecology, a new interdisciplinary paradigm, introducing sustainable industrial ecosystems through interaction between industrial symbiotic networks and their environment. Industrial symbiosis is also understood as part of industrial ecology (Chertow 2007; Rui and Heijungs 2010; Howard-Grenville and Paquin 2008; Gingrich 2012 in Rončević and Fric 2015, pp. 35–36) or as a possible approach to industrial ecology. One possible difference between industrial ecology and industrial symbiosis is in their focus; while industrial ecology is principally concerned with the flow of materials and energy through systems at different scales, from products to factories and up to national and global levels, industrial symbiosis research is more focused on flows through geographically proximate networks of businesses and other organizations in local and regional economies. Industrial symbiosis also sometimes appears as a synonym for industrial ecology (Phillips et al. in Rončević and Fric 2015, pp. 35–36). Some authors avoid this conceptual confusion by focusing on industrial symbiotic networks.

Industrial symbiosis represents a relatively new interdisciplinary research area, combining research in its technological, ecological, informational, economical and since recently also sociological aspects. This interdisciplinary research is becoming increasingly relevant due to the fact that intentional or planned development of industrial symbiotic networks has immense potential due to their contribution to the reduction of entropy via saving in energy and materials and reduced waste production, and hence its contribution to increase in value-added and quality of life. These results are achieved by improving environmental performance of industry, by increasing effectiveness of technologies and processes for the reduction of environmental damage and especially by connecting flows of materials and energy, in the framework of industrial symbiotic networks.

These networks can be defined as a collective system of flows connecting a multitude of actors in the framework of a geographically proximate area (e.g. micro-region, industrial district), who coordinate exchange of resources in pursuit of their own business interest (rational action), but which leads to the above-mentioned collateral benefits. By working together, businesses ensure not only their economic benefit, but attain collective benefits that are greater than the sum of individual benefits achieved from acting alone. Positive economic effects are demonstrated in lower waste/by-product disposal costs, competitive advantages towards other economic actors and better reputation of the company/economic actor. Industrial symbiotic networks therefore exist for the purpose of an exchange of resources, such as by-products, materials, services, energy, etc., in the framework of closed and sustainable cycles. This cooperation contributes to the reduction of required input of materials and energy, as well as the reduction of the output in the form of waste (Chertow and Lombardi 2005). Furthermore, the low price of waste as a material - if compared with non-waste materials – increases the economic efficiency of individual organizations, which are recipients in the framework of industrial symbiotic networks (ISNs). Open networks, on the other hand, allow entropy, which results in the loss of energy, poor utilization of materials and consequently cause environmental damage. However, although the formation of industrial symbiotic networks seems rational both from an organizational and from a societal point of view, and although we have sufficient technological and logistical knowledge to manage them successfully, industrial symbiotic networks are poorly developed in most environments: they include a small number of actors, are not robust enough to include a relevant proportion of flows of resources and are not recognized as relevant by the policy makers.

2. How to tackle ISNs as a social phenomenon? Outlining a research program

Why is this the case? We have already stated that industrial symbiotic networks transcend the boundaries of various systems, which includes the social system. This implies that they are not only technological networks, but are also social networks. Any comprehensive research effort should therefore explore the industrial symbiotic networks, their emergence and evolution also from the perspective of social science. Nodes in industrial symbiotic networks – enterprises, households, policy actors - are social actors, operating in the context of bounded rationality, i.e. they are limited in their rational decision making not only by their own inherent organizational values, but also constrained by the decisions of other organizations. Since the exchange - the elementary activity in industrial symbiotic networks – is taking place between social actors, we can hypothesize that industrial symbiotic networks are also social networks and their morphogenesis can be studies by sociological research. An exchange requires cooperation, communication, and a certain level of trust. Other researchers have already suggested that the effectiveness of industrial symbiosis depends strongly on trust and reciprocity, i.e. cooperation between economic actors involved in industrial symbiosis (Domenech 2010, pp. 81–282). Hence, research on industrial symbiosis also requires us to explore its social aspects (Rončević and Fric 2015).

Tackling the social aspects of ISNs and the mutual interplay with other dimensions requires a new research perspective. In view of current research, which is significantly underdeveloped, we should not stay short of proposing a new research programme. This research programme would have to:

1. Develop robust input-output model of physical, informational and social exchanges in ISNs, including their social interactions (networking); the ultimate

aim is to model morphogenesis of ISNs by taking all relevant aspects into account. We will engage it in computational sociology to explore, model, simulate, test and observe ISNs.
2. This model should allow us to predict morphogenesis in a specific social setting, thereby allowing us to develop tools for embedding industrial symbiosis in its natural, technical and socio-cultural setting.
3. In this interdisciplinary research, one should show connections and the exchange between dynamic, completive, adaptive networks, operating across the boundaries of technical, natural, computational/informational and social systems.
4. With the model one should identify the prerequisites and mechanisms for successful morphogenesis of industrial symbiotic networks, which will enable us to highlight already existing potentials and barriers to industrial symbiosis within a specific setting. This includes the factors which are necessary to improve collaboration among actors;
5. This research will focus on social aspects of industrial symbiotic networks. Hence, it is necessary to explore the contemporary sociological literature and theories of social action (Bashkar, Mouzelis, Bourdieu, Giddens, Archer, Sztompka, etc.) and use these theories to conceptualize the morphogenesis of symbiotic networks; we will also examine the influence of social structures on individual actions of ISNs members.

Proposed research programme should search for answers to the following key research question: *What are the factors that influence the emergence and evolution of industrial symbiotic networks and what are the underlying mechanisms?* Answering this question will, first, allow us to understand whether and how we can effectively influence the development of ISNs with purposeful action. What key elements can enable the emergence of industrial symbiotic networks and their interactions? Secondly, we will be able to explore not only the explanatory role of social theory, but also its applied potential, theoretically inspiring morphogenesis of ISNs. Thirdly, we will be able to explore one of the basic sociological questions, namely the mutual influence of emergent social structures and individual actions (Archer 2003). Fourthly, we will be able to study and explain substantial differences in the morphogenesis of ISNs in different socio-cultural settings. With the development of information and communication technologies, numerous formal and informal networks started to emerge. Organized networks of structures and organized networks of actors are referred to as collaborative networked ISNs. In industry, clusters of organizations aim at efficient product development, production, and marketing tasks. Such partner consortia function as networked organizations in which cooperation is supported by information and communication technologies. Mutual trust can contribute to improved knowledge sharing, resource sharing, and taking joint risks. Trust brings many advantages to ISNs and should, therefore, be considered beforehand when creating ISNs.

While answering the research question, the proposed research programme should test the hypothesis that industrial symbiosis is more than environmentally sustainable industrial activity. It is not only a paradigm with important technological, economic and environmental consequences, but is also a social paradigm and its manifestations in the form of industrial symbiotic networks are also social networks.

3. Research Methods and Industrial Symbiotic Networks

A preliminary literature overview shows that even though the area has attracted much academic attention and has been reported to lead to economic and environmental benefits (Chertow and Lombardi 2005), most of the initial contributions focused on the engineering and technical feasibility of the exchanges, whereas social elements remained mostly unaddressed. Although relevant literature has partly addressed this gap and recognized the role played by social aspects, there is still little understanding of how social mechanisms work; how they affect the emergence and operation of industrial symbiotic networks and, most importantly, there is a lack of comprehensive frameworks for the analysis of the soft elements of IS. Some publications offer conceptual frameworks or theories of a normative nature. These are designed and applied to prescribe how industrial symbiotic networks should develop but they fail to properly analyze and explain specific empirical cases or allow generalizability of results.

The interdisciplinary research program should strive to capture the physical, organizational and most importantly the social aspects that surround the process of emergence and development of industrial symbiotic networks. The research focuses on the understanding of the social processes that govern material and energy exchanges. Therefore, due to the nature of the research questions, the methodology design should take a predominantly inductive approach, using a combination of qualitative and quantitative methods (method of triangulation). The research should depart from a fundamentally constructivist-interpretationist paradigm.

In selecting the appropriate research approach we will have to answer some basic ontological and epistemological questions, concerned with identifying, in the most general terms, the kinds of things that actually exist in industrial symbiotic networks. In other words, addressing the question: What is the nature of existence of industrial symbiotic networks? We have to be concerned with the nature of knowledge itself, its possibility, scope, and general basis; more broadly: How do we go about knowing things around industrial symbiotic networks? How do we separate true ideas from false ideas and representations? What are the systematic ways we can determine when something is functioning effectively or not? So, ontological and epistemological premises guide not only the selection of the research focus, but also the methodology adopted and the expected outcomes. Industrial symbiotic networks are thus examined as a social structure that is simultaneously constructed and reflected upon by the researcher and actors involved – this approach understands their identification first as a singular phenomena, and then as a sustainable instrument, part of the process of co-creation of meaning and social interaction. This ontological and epistemological position explains the focus of the research question on the social processes behind the articulation of industrial symbiotic networks and the emphasis given to structural and discursive dimensions in the analysis.

In this research process, the qualitative data should indicate the research questions that will be outlined. While selecting relevant data, the aforementioned ontological and epistemological issues of the research should be properly addressed. The process of selection (based on physical, organizational and social conditions) of case studies, the criteria used and the decisions of appropriateness of industrial symbiotic networks require clear identification of the criteria, i.e. attributes of ISNs. These can help distinguish successful from unsuccessful performance of industrial symbiotic networks and use only reliable and valid predictive measures of success. Not least, the main limitations of proposed research program should be carefully scrutinized.

In our own attempt, which will fall within the proposed research program, basic concepts required for modelling sociological phenomena will be used for modelling and the data will be collected from secondary empirical case studies. We will lean upon our own multi-agent system architecture combined with extensions for social modelling. While sociologists often model social processes as interactions among variables, we review an alternative approach that models social life as interactions among adaptive structures and agents who influence one another in response to their influence. These so called agent-based models (ABMs) show how simple and predictable local interactions can generate familiar but enigmatic global patterns, such as the diffusion of information, emergence of norms, coordination of conventions, or participation in collective action. Emergent social patterns can also appear unexpectedly and then just as dramatically transform or disappear. We will discuss how theoretical interest focuses on dynamic social networks that shape and are shaped by agent interaction. ABMs will be used to perform virtual experiments that test micro-macro sociological theories by manipulating structural factors like network topology, social stratification, etc. The designed framework therefore builds on social theories and computational sociology. We introduce three models to support the decision makers in the process of industrial symbiotic networks. In particular, we propose models for physical, organizational and social aspects of industrial symbiotic networks. Then we will build our new model.

We should emphasize that social aspects are the key added value. Establishing trust is one of the most important mechanisms in collaborative projects, since the possibility of opportunistic behavior of partners cannot be eliminated by formal contracts. In moderately sized industrial symbiotic networks of a few dozen partners, means for trust building include personal meetings, regular communication, sharing of information and knowledge, and stable rules of the game, whereas computational means can also be beneficial to support trust modeling (Klein and Methlie 1995). Computational means for trust modeling are, however, indispensable in large IS networks in which personal meetings and communication are hindered by the size of the IS. In order to propose a mechanism appropriate for modeling trust between collaborating partners, the conditions for trust building and management need to be analyzed first.

The construction of any model within this research program will be possible by following a meta-analysis method to "guesstimate" cause-effect relationships, in regards to a critical-realistic philosophical perspective. When trying to construct a model and "reduce" a perceived reality of IS systems, the research encompasses and gives importance to social interactions. These interactions are assumed to constitute the grounds of synergies between nodes in ISN social network. This will be undertaken at two different levels: (1) in the methods of data collection by using other sources of information, mainly secondary material about the network and actors and, (2) in the preparation and selection, by contrasting and cross-comparing the secondary information that will be put into software.

We will choose Decision Support Methodology, as it is an approach that relies on an interaction study with some main industrial symbiosis case studies and information is continuously reinterpreted in an iterative process of collection and analysis of data. This methodology is a broad discipline concerned with supporting people – e.g. policy-makers – in making decisions. It is a part of decision sciences, together with normative and descriptive approaches to decision making (Clemen 1996). Decision support encompasses a number of disciplines, including operations research, decision analysis, decision support systems, data warehousing, and group decision support (Jereb et al. 2003). The general approach originates in decision analysis, a discipline popularly known as applied decision theory. Decision analysis provides a framework for analyzing decision problems by Clemen (1996): (i) structuring and breaking them down into more manageable parts; (ii) explicitly considering the possible alternatives, available information, involved uncertainties, and relevant preferences; (iii) combining these to arrive at optimal or at least "sufficiently good" decision.

Nevertheless, we should especially emphasize that other analytical approaches are possible and should be taken into account.

4. Main benefits of proposed research program

Proposed research program should explain why, under similar material conditions, the actual emergence and operation of ISNs can result in radically different outcomes depending on the socio-institutional framework in which they operate. Different levels of analysis can be included: the structure and morphology of the network, social dynamics, social processes, the underlying environmental discourse and the development of methodology, by using a simulation - creating a model of ISNs with the scope to find key features of such ISNs.

An analytical study and its application to different typologies of networks is trying to understand how can all these social processes foster or hinder the further implementation of ISNs in some area. Sociological approaches to the IS may identify three types of social structures relevant for the explanation of outcomes: social networks, institutions, and cognitive frames. In this analytic context, it is critically discussed the varied notion of a middle position on the ongoing theory between positivism and constructivism, as well as the epistemologically beneficial role that meta-theoretical reflexivity and the internal conversation can potentially play (Beckert 1999). Observation and testing of theoretical cause-effect, not only social relationships but also physical and organizational, have been possible by the use of computational simulation. However, other theoretical perspectives, which may target other explanations, are possible as well.

The ISNs concept and its characteristics based on environmental, organizational and technological levels can be supported by applying sociological theories, i.e. interrelations between social structures and agents, as discussed by sociologists for example Archer, Giddens, Sztompka, Bhaskar, Mouzelis, Bourdieu, Vandenberghe. The micro-macro or agency-structure questions are indisputably one of the most important theoretical issues within the human and social sciences (Bhaskar 1989). The social-theoretical dimension of ISNs adopted here is relevant from both perspectives: the circularity and the relative autonomy of structures vs. actors, or of institutions vs. individuals (Giddens 1994).

One of the key foci of the proposed research program should be modelling of social interactions. Usually, models of social interactions begin with a framework of how an individual is affected by others in his reference group. The reference group is simply the set of other agents whose behavior the focal agent can be affected by. This definition is broad, and can include all other agents in the economy ("macro"-level models), or small social groups ("micro"-level models). Actors in ISNs have stable expectations and are formed by the structural, institutional and cultural embeddedness of (market) exchange (Archer 2003). This reveals ISNs exchange to be a form of social interaction that cannot be explained by a "natural propensity to truck, barter and exchange" Smith (1976) but only by the institutional structures, social networks and cognitive frames of meaning within which actors inside ISNs meet.

We should emphasize that without a systematic approach that enables reliability and validity, decisions of ISN members (companies) and policy-makers remain subjective, of dubious value and open to challenge. In contrast, the robust models with a strong relationship between predictors and criteria, lead to better informed decisions and allocation of resources. In such a case it is easier to satisfy requirements that analytical selection procedures must be objective, nondiscriminatory, and result in the best solutions being selected (Stone 1998). The choice of selection criteria (attributes) should be consistent with the organization's strategic direction and culture (Stone 1998). Finally, as Cascio (1991) points out, "more accurate predictions result in greater cost savings (monetary as well as social)". This selected approach to support the selection of appropriates of existing ISNs features generated from several IS cases around the world, integrates multi-attribute modeling and expert systems. The basic principle of the multi-attribute decision making is a decomposition of the decision problem into smaller less complex sub-problems. This makes the problem transparent, and understandable; the decision explainable, and more effective (Jereb et al. 2003).

Trying to construct robust models and "reducing" a perceived reality of ISNs, the research work will give appropriate attention to social interactions in technological networks. These interactions constitute the grounds of synergies between companies. These models would allow evaluating, confirming and spotting the morphogenesis of these networks and their effects, as well as comparisons of the presumed interactions and their mechanisms in different social settings, and also potentials and barriers to the development of ISNs. They will allow evaluating, according to the models, the existence of necessary and sufficient conditions of successful industrial symbiotic networks. Ultimately, this will (1) enable ISN members and policy-makers to make informed decisions based on results of scientific research; (2) provide measurement of ISN evolution and performance; and (3) show the presence of desired causes. Not least, it will (4) add an interesting research problem, enabling development of a newly emerging sub-discipline of computational sociology. Computational sociology will help in developing and testing theories of complex social processes through bottom-up modeling of social interactions. This implies bottom-up modeling that involves the understanding of social agents and structures, the interaction among these two, and the effect of these interactions on the social aggregate.

5. Conclusion

We believe that the proposed research program is comprehensive and has the potential to contribute not only to fully appreciate and understand morphogenesis of ISNs in their social context, but can genuinely contribute to further development of the nascent and fledgling field of exploration. It will also have a number of applied dimensions, such as:

- Applying new feasibility models in order to reveal the already-in-place potentials and barriers to the development of ISNs.
- It will allow evaluating how well the preconditions for a successful ISN morphogenesis and its efficient operation are fulfilled in a specific social setting.
- It will allow qualitative assessment of perception of each other businesses (business transparency issues), awareness of environmental pressure and IS networking benefits.
- Carry out assessment of social interaction in the framework of industrial symbiotic networks, including micro-macro level integration. Social structures and agent-based modelling are best suitable to show the emergence of patterns from individual decisions and social interactions.
- By introducing these models, we will be able to explain and possibly forecast new ISNs and their behavior.

6. References

Archer, Margaret: "*Structure, Agency and Internal Conversation*". Cambridge: Cambridge University Press, 2003.

Bhaskar, Roy: "*Reclaiming Reality: A Critical Introduction to Contemporary Philosophy*". Verso: London, 1989.

Beckert, Jens: "Agency, Entrepreneurs and Institutional Change: The Role of Strategic Choice and Institutionalized Practices in Organizations". *Organization Studies 20*, pp. 777–799, retrieved 26.5.2015, from http://oss.sagepub.com/cgi/content/abstract/31/5/605.

Cascio, Wayne: "*Applied Psychology in Personnel Management*". Engelwood Cliffs: Prentice-Hall, 1991.

Chertow, Marian R. / Lombardi, D. Rachel: "Quantifying Economic and Environmental Benefits of Co-located Firms". *Environmental Science and Technology*, 39(17), 2005, pp. 6535–6541.

Clemen, Robert T.: "*Making Hard Decisions: An Introduction to Decision Analysis*". Duxbury Press: Boston, 1996.

Domenech Aparisi, Teresa Ana: "*Social Aspects of Industrial Symbiosis Networks*" (Bartlett school of Graduate Studies: University College London). (doctoral thesis) 2010, pp. 81–282.

Giddens, Anthony: "*Living in a Post Traditional Society. Reflexive Modernization*". Polity: Cambridge, 1994.

Jereb, Eva / Bohanec, Marko / Rajkovic, Vladislav: "*DEXi – Računalniški program za večparametrso odločanje [DEXi: Computer Program for Multi-attribute Decision Making]*". Moderna organizacija: Kranj, 2003.

Klein, Michael / Methlie, Leif B.: "*Expert Systems: A Decision Support Approach with Application in Management and Finance*". Old Tappan, NJ: Addison-Wesley, 1995.

Phillips, Paul / S. Barnes, Richard / P. Bates, Margaret / Coskeran, Thomas: "A Critical Appraisal of an UK County Waste Minimization Programme: The Requirement for Regional Facilitated Development of Industrial Symbiosis/Ecology". *Resources, Conservation and Recycling,* Elsevier: Melbourne, 46(2005), pp. 242.

Rončević, Borut / Fric, Urška: "Researching industrial symbiosis: challenges and dilemmas". In: Povh, Janez (Eds.). *Applied Modelling and Computing in Social Science*. Frankfurt am Main: Peter Lang 2015, pp. 35–49.

Smith, Adam: "*The Wealth of Nations.* Chicago". University of Chicago Press, 1976.

Stinchcombe, Arthur L.: "The Conditions of Fruitfulness of Theorizing about Mechanisms in Social Science". *Philosophy of the Social Sciences*, 21(3): 1991, pp. 367–388.

Stone, Raymond J.: "*Human Resource Management*". New York: John Wiley & Sons, 1998.

Lea Prijon
School of Advanced Social Studies in Nova Gorica
Gregorčičeva 19, 5000 Nova Gorica, Slovenia
lea.prijon@fuds.si

Networking of Business Elite in the Information Society: The Case of Interlocking Directorates

Abstract: *Contemporary business elite gain and retain their privileged position, power and ability to control society through merging in (in)formal networks, as one of the crucial characteristics of the information society. The latter is linked with the network society and networking as one of the key element of the new economy. The information society and new economy are the result of globalisation, therefore we speak about the decentralised phenomenon, which needs a linker in order to operate successfully. This linkers are the so-called business elite, who merge in interlocking directorates, which are reserved for privileged individuals in order to gain power and impact society. The latter is formed through strong networks (ties) in the inner circle in interlocking directorates, which provide access to relevant business-related information for important business transactions, which enable significant economic and non-economic profits for the economic elite. In this sense, the business elite act as crucial (decision making) actors in the globalised information society.*

Keywords: *business elite, information society, network society, information, social capital, power, interlocking directorates*

1. Introduction

The information society is characterized by numerous elements, among which the most important ones are precisely Information and Communisation Technologies (ICT) and information, as a product of ICT. Information is the key prerequisite and source for a privileged status in society and power, used by individuals to constantly and importantly impact society and its subsystems, thus having and maintaining a privileged social role.

In the present article, we focus on the analysis of the business elite in the information society, while also the elite's position and influence, which is ensured and enhanced with and through integration and merging in formal and informal networks. Merging is crucial for the business elites' existence, which act as an organized group of individuals in a dispersed and decentralized Information society. Through merging social capital increases, while social ties strengthen, which facilitates and increases possibilities for accessing the inner circle and important sources of business-related information. In the present article, the business elite is considered as a sub-segment of economic elite, thus a group of people who occupy strategic positions in important corporations, which enables them the access and the use of social power. In addition to strategic positions, ownership of (non) economic resources is also of vital importance, among which information is one of the most important in the information society. In fact, information allows businessmen the realisation of significant business transactions, which help consolidate their already privileged position. On the other hand, the absence of information could seal the collapse of business and thus power of specific businessmen. The latter reflects the importance of business elite's merging, as their (business) success depends on mutual integration and collaboration. We specifically focus on the analysis of interlocking directorates, which are the result of business elites' integration, represented by boards of directors. Boards of directors are interlinked and have a common director, who can simultaneously serve as a director of two or more corporations and represents the primary decision-making body. Interlocking directorates emerged in '70, but their structure and impact in the (global) economy has changed. Already at its beginnings such associations were important for promoting the upper class cohesion, facilitating business transactions, etc. In fact, interlocking directorates remain an important tool to facilitate business and a crucial means for exercising social power and impact in the information society.

As such, interlocking directorates represent the "*basis and source*" of crucial business information and therefore an important form of the business elite's networking. Despite the fact that in its early stages of emergence, interlocks have been labelled as harmful and negative and as a means, which enabled and promoted harmful clustering as an obstacle to competition, and despite numerous acts which have been adopted to regulate such merging, this kind of integration still remains a popular form of networking for the business elite. However, the purpose and operation of contemporary interlocks is much different, as they are controlled by legal acts. Nevertheless, interlocks still facilitate strategic positions, power and possibilities for control, which have become crucial non-economic sources through and within which members of the business elite exercise their social power and control social (economic, political and cultural) processes.

2. What is the Information Society

The information society, which is also generally used synonymously with Bell's (1973) post-industrial society, is becoming a widely spread and used word to describe modern society, which is inevitably linked with globalisation, as a driving force of development. "*Globalisation in this context means the creation of centrally coordinated, globally distributed networks of producers and consumers, supply chains, and distribution networks*" (Gurstein 2008, p. 9), which was further empowered by Information and Communisation Technologies (ICT). In fact, new technologies are one of the crucial and most obvious indicators of the new era, which distinguishes it from industrial society. Nevertheless, there are also other indicators that suggest the existence of the information society, such as occupational changes (characterized by the increase in jobs for white collars), new economic principles (primary and secondary information sectors of the economy), spatial changes (information networks as prominent features of social organization) and cultural changes (as contemporary culture is information led and takes place in the media-saturated environment) (Webster 2014).

Technology infrastructure enables local and dispersed entities to efficiently perform their own responsibilities and tasks in order to achieve centrally established aims, as well as respond to external standards. This defines technology as "enabling force", more than a repressive or controlling one (Gurstein 2008). In this respect, ICT can be defined as a revolutionary tool, which has changed the way of working, transformed politics, by promoting democratization (Webester 2006) and economy, by promoting economic success, it affected people's lifestyle, norms, values, way of thinking, etc.

Between 1996 and 1998, Castells published a three-volume study: *The Information Age*, which had a great impact on contemporary scientists' thinking, as he analysed the role of information in a contemporary period. Castells (2004) claimed that the Information Age was to announce a *New Society*[1] characterized by the development of networks (enabled by ICTs), which foregrounds the flow of information, leading to the so-called phenomenon of the *explosion of information*. To this extent, Webster (2006, p. 26) defines information in the basis of semantic definitions, claiming that: "*Information is meaningful; it has a subject; it is intelligence or instruction about something or someone*" (Webster ibid). In addition, information is used as an economic resource by economic organizations, which increase efficiency, innovation, competitiveness, etc., while also as an aspect of economic, social, cultural and political life (Moore 1997). Moreover, Castells

1 At this point not yet anticipating the *information society (or era)*.

(1996) speaks about the so-called "*network society*", which arose with the merging of capitalism and information revolution, and is not just a result of globalisation. The network society has had an important impact on societies' organisational forms since the occurrence of global de-bureaucratisation of affairs, and it can been defined as "*the organizational arrangements of humans in relationships of production/consumption, experience, and power, as expressed in meaningful interaction framed by culture*" (Castells 2000, p. 5). Therefore, the information society and network society have become related concepts and synonyms for the contemporary society in which networking (merging) represents a crucial means of social power and influence.

The information society and network society originate from long-term economic developments and technological changes (Moore 1997) and form the so-called "*new economy*", which is *informational* as it generates knowledge, while information determines productivity and competitiveness of economic units. The new economy occurs in an interconnected and mutually dependent space, which allows participation of economies of different countries and is therefore a *global phenomenon*. Despite the fact that the new economy involves different countries' markets, it still operates as a mechanism that connects various units, therefore it is *networked*. It can be defined as a new form of economic organization, called the *network enterprise* consisting of various enterprises or segments, large corporations, etc. which are internally de-centralized as networks, while small and medium businesses are joined in networks. These links are specific as they connect among themselves (e.g. business projects), while this kind of cooperation is based on sharing, acquisition and exploitation of information, through the *information network* (Castells 2000, pp. 10–11). The network society is also characterized by a rapid response and adaptability on the global market, therefore networks are the crucial element for this kind of society to operate and retain. In essence, corporations have changed into networks, operating under strategic alliances (Castells 1996).

Starting from the latter, in the present article, we will analyse networking in the information society, within the context of the new economy, with a special focus on interlocking directorates as a special form of the business elite's merging. We will derive from Castell's (2004) conceptualisations of networks as a "*set of interconnected nodes*", which are dispersed (without a centre) and which importance increases in accordance with the meaning of information that these networks contain, and in accordance of their goals. Nevertheless, all network's nodes are crucial for its effective performance, which can also work as structures of power and as a locus of decision making bodies (Szalacha 2011). Therefore, we derive from *information* being the crucial *capital* in the information society and for centres of power to emerge. In addition, despite the explosion of information, the latter remain(s) a very valuable prerequisite for success, innovation, speed product development, productivity, competitiveness, etc., if properly managed and used (Moore 1997). Nevertheless, the nature of information remains difficult, even mysterious to understand as it is intangible, non excludable and crucial for contemporary market dynamics. It is also true that information is a "fluid source", once it is produced or known, it is almost impossible to prevent its dispersion (Castells 2000). It is because of the latter, that information are of vital importance and a valuable resource(s) in the new economy, since on one hand it is difficult to prevent its occurrence in public, while on the other it is crucial to find the right information in the information explosion and take advantage of the latter to gain benefits. Besides, as Moore (1997) already states there are many factors, which can reduce the access to information for general citizens, which can cause inequalities among individuals (elite vs. masses).

It is precisely for this reason that networks are so important for the business elite, as these facilitate the access to information and thus work as a source of power. In what follows we focus on a specific type of networks, namely *interlocking directorates* in which the business elite join in order to retain and gain power in a legitimate way and consolidate power, influence and existence.

3. Business Elites and Centres of Power: Strategic Elite and Power Elite

Society is a social structure "*formed by the interplay between relationships of production/consumption; relationships of experience; and relationships of power*" (Castells 2000, p. 7). *Meaning*[2] is (re)produced through symbolic interactions between actors, who form social structure and "fight" over power (Castells ibid.) and opportunities to govern society, therefore individuals occupy different social positions and have different degrees and abilities to influence society. Due to the fact that the information society is decentralised since nodes (within networks) are dispersed, there is a need for a leading source or a linker, which will be able to influence and coordinate societies' processes in order to work effectively. These groups are the *elites*, which were defined by Pareto (1935) as a privileged group of individuals, who represent the top of society and have the highest index of activity, which derives from unequal distribution of talents among individuals.

2 "Meaning" is understood by Castells as symbolic identification by an actor of the purpose of her/his/their action.

Contemporary elites[3] do not represent a cohesive whole (Tomšič 2011), but dispersed centres of power, who cooperate and merge when it comes to the issue of their existence, benefits, influence, impact, etc. As already mentioned, elites hold strategic positions in important organizations and hold (non)economic means, which allow them to substantially affect political (Higley 2010) and economic outcomes, therefore they have levers for social power. As the article deals with the merging of the business elite, as a sub-segment of the economic elite, the latter is understood as a group of individuals within the aspect of economy, who poses and/or controls (non)economic resources (Mills 1965), such as money, means of production, shares, gold, while also information. When defining the business elite we can relate and derive from the economic elite, which is a group of individuals who control economic processes and have the highest concentration of (physical/economic) capital and income in form of profits, which are reinvested or spent. The business elite owns means of production, economic resources, manages economic processes, possess great concentration of capital, etc. (Figueroa 2002, p. 8). Therefore, we speak about individuals, who occupy leading positions in strategic organizations or corporations and hold positions that enable power and control. The latter enables this group of people to constantly and significantly affect the income of other individuals in society (Figueroa 2008).

Mills (1965) considered the economic elite as a part of the *power elite*, which implies the existence of a close symbiotic link between policy (political elite) and entrepreneurs (managers) and business owners. In this context the business elite can be considered as a part of the power elite, since it consists of individuals who are located in strategic positions, which allow them to make decisions with great influence on society. The economic elite are individuals with the largest property and who occupy key positions in large and important economic organisations, which are a prerequisite for power, wealth and social prestige and at the same time, the main instrument of authority, as well for acquisition, retention and consumption of higher forms of social prestige (Mills 1965, p. 101).

Within classical definitions of elite, Mills (1965) argues that the individual's initial position in the stratification system primarily depends on its economic and/or non-economic resources, but also adds that elite are not (necessarily) the

---

3 Generally speaking, there are three segments of elites; political, cultural and economic (business elite can be considered as it's sub-segment), which are linked with criteria of power, knowledge or skills as their key attributes. Members of each elite's segment hold crucial (non) economic resources and have a common interest, i.e. *maintain their privileged position*, which creates tensions and conflicts between these three segments, while also within the specific segment itself.

richest people in society. In fact, in addition to (non) economic resources, an individual has to occupy strategic positions in important and influential corporations, in order to be classified as an elite. The latter is a prerequisite for social power, wealth and prestige and, at the same time, the main instrument for acquisition and retention of authority and sources of impact. However, since an individual can improve (or lose) its position over time, by virtue of his status, property and other (non) economic resources, the elite's status is fluid, flexible and variable in time and space. In this respect, the contemporary business elite form the so-called strategic elites (for more on strategic elites see Keller 1991; Etzioni-Halévy, 1993), which characterize modern social order, which is moving from a *hierarchical organization* into *stratarchy*, and implies the existence of different sectorial pyramids, governed by various elites, which are mutually separated and their work is uncoordinated.

Taking into account the perspective of elites as "*individuals who are able to constantly and significantly affect national policy due to their strategic positions in major organizations*" (Field et al. 1990, p. 152), the strategic elites (the power elite) can be defined as key actors in decision-making processes, who can influence all social subsystems. However, these elites do not constitute a single corpus of power (Tomšič 2011), but act as relatively autonomous groups of individuals, who do not work with a common strategy but do pursue a common goal, i.e. *accumulation of social power and prestige*. As elite groups include only privileged individuals, who occupy strategic positions, they constitute a rather small group of people, who act in decentralised and disperse society. Therefore, it is crucial for them to join (merge) in (in)formal networks in order to gain the power to rule and control, the masses, as well as to gain other (non) economic resources and prestige.

In the context of contemporary elites, the latter achieve their goals by cooperating and merging in order to achieve important business-related information, which will allow important business transactions and consequently strengthen their status, wealth and privileged position. This is even more important for business elites, which act in the turbulent and ruthless global market, hence networking has become the only possible means to maintain their privileged position. They gather in specific networks, called interlocking directorates, which enable important business transactions, but are still legally controlled (e.g. *The Sherman Antitrust Act* 1890; *The Clayton Antitrust Act* 1914) in order to ban the formation of trusts, various restrictions of competition, etc.

In the following paragraph, we analyse the process and meaning of elites' merging in networks, with a special focus on interlocking directorates, which represent the key source of power for the contemporary business elite. Interlocks allow business elites (considered within the context of power elite and strategic elite) to gain information, through merging in (in)formal networks as one of the key sources of power in the information/network society. At this point, it is inevitable to mention social capital as one of the crucial means for the empowerment of the business elites, as it is formed with and within merging. Social capital is an intangible source, but yet one of the most important means for facilitating privilege and power positions in the networking society and the era of new economy.

4. Merging of Business Elites and Social Capital

When we research networks and networking, we cannot bypass the phenomenon of *social capital*, which is formed with integrating and connecting in networks. It is not produced by itself, as individuals have to "invest in it" (by merging), while at the same time its intensity rises when interpersonal relationships are getting stronger. Social capital is extremely valuable and important as it enables the acquisition of resources such as money, power (Adler and Kwon 2000) and information. Social capital is a social resource, which refers to connections within and between social networks and as such represents collective *goods*. It represents individual's relationships with other members of society, it grows and is maintained through creating new links (Portes 1998).

Social capital is the aggregate of actual and potential resources, which are owned by durable networks maintained through mutual acquaintances. These are characterized by the presence of relevant sources of social networks, availability of resources to individuals who are included in networks and utility of resources for individuals and groups (Bourdieu 1986). It can be defined as an intangible phenomenon, which acts as the facilitator for integration in the environment by establishing channels of communication, thereby promoting developmental performance (Adam et al. 2005). It exists in limited quantities and represents a combination of three network properties, i.e. *the presence of significant resources* (contained in a particular social network), *the availability of these resources for members of network*, while also the *use of these resources for individuals and groups in terms of effectiveness and efficiency* (in terms of the objectives and costs) (Bourdieu 1986).

Social capital is formed through networking and is therefore a result of *social ties*, which are of vital importance for new opportunities, success in business and integration in social networks as they determine the degree of integration(s) and thus the amount of social capital and power that individuals may obtain through networking or within the network. This kind of networking and social capital includes horizontal and vertical forms of association and behaviour, within and between different structures (e.g. enterprises) and entities (individuals). Vertical networking represents hierarchical links, and uneven distribution of power, which is reflected in the unequal distribution of resources, rewards, power, control options, etc. Therefore it is essential that individuals, who want to gain money, power, information and other benefits, are integrated in networks and establish powerful ties in order to raise their opportunities for benefits. This is the logic of business elites, who, by reinforcing their social capital strengthen their power, position and impact in society and further retain their privileged position.

As already Castells (1996) notes, networking has become an important element of the information society, as it goes hand in hand with the network society. The fact is that networking becomes a key component of the contemporary society for all individuals, nevertheless the power of their networks and ties, while also social capital, which forms within these networks, differs greatly from the elites' networks and ties. The business elite merge in specific forms of networks, which ensures the access to important sources of business-related information that allow them important and crucial business transactions, on which their existence (as important and influential businessmen) may depend on. Therefore, members of the business elite merge in interlocking directorates.

In this context, sociologists and experts have different views on the elite's networking/merging. Starting with pluralist democratic experts (e.g. Dahl 1957; Parsons 1960), who derive from conceptualisations that Western social order is characterized by dissociation and diversification of power (polyarchy), which enables the elite's competition, and implies the necessity of controlling the elite's power by democracy itself, believe that competition and integration of the elites in democracies does not have a strong impact on the economy. The aforementioned authors believe that especially in nondemocratic regimes, there is collusion among the elites who hold political and economic power and act in accordance with their own interests.

On the other hand, representatives of the classical elite's theories (e.g. Mosca 1939; Pareto 1935; Mills 1965) define the elite's integration and merging as a process of concentration of power, even in democratic regimes. In these cases, the elites do not compete for power and authority and do not control and balance each other, but are linked in an unanimous cohesive power elite (Domhoff 1970).

Figueroa (2002) and Mills (1965) argue that the business elite cooperates with the political one, and vice versa, for a successful and long-term preservation of privileged position and status. At this point, it is crucial to add that the long-term existence of the business elite and preservation of power also depends on the level of integration within the same business elite and their level of integration in the economic sphere in general (various economic institutions). Therefore, merging and networking inevitably leads to the formation of the so-called power elite. Besides, organized interest groups are more influential and powerful than an individual (business) elite, while also business-oriented groups have a strategic role (Giles and Page 2014).

Merging in (in)formal networks facilitates the business elite market's monitoring and competition and can emerge due to long-or short-term interests, which act as a positive or negative component, with different dimensions and impact on society (for more on networks see Prijon 2010; 2015a; 2015b, Prijon and Prijon 2015). Aggregation is a characteristic connecting element of the elites, which occurs in all economic environments associated with investments and transfer of capital. In addition to typical social integration characteristic for strong ties (Granovetter 1983; 1985), integration towards profit-non-profit is usually supported with integration in direction of money-power, typical for all forms of elite structures, both in democracies and authoritarian regimes.

Networking occurs in different ways and for different reasons, nevertheless its common point is facilitated access to various resources that enable elite's power, impact, importance, benefits, etc. Networks are characterized by relations of reciprocity and amenities which generate economic returns through recognisability of the economic player on the market and its trustworthiness for achieving funds, credit insurance, etc., whereby social networks lower cost of such transactions. In addition, economic returns are generated also through social networks, which protect individuals in the event of losses which may result from uninsured market risks (such as moral hazard) that may cause bankruptcy or other economic catastrophes (Figueroa 2002, p. 10).

5. Business Elite and Interlocking Directorates

The new economy, as a result of changes and developments in the information society has brought important changes also in the economic sphere and business sector in general. One of these changes was a shift from family capitalism through managerial capitalism to institutional capitalism. The crucial characteristic of such an economic arrangement are impersonal corporations, which are mutually connected by strong ties than ever before, which arose from the so-called *inner circle* of interlocking directorates, as business formations typical for modern capitalism (Useem 1984). The role and functioning of interlocks has changed from the time they first started to appear in their primitive forms (as cartels, trusts, etc.), as a result of an increased regulation of the latter. A key change occurred as they have become more integrated and connected due to larger and more powerful interdependency for business (un)successfulness, which is a result of the decentralised information era (society). Businessmen merge in order to successfully operate as players in the global market and in order to become stronger business players. In this matter, contemporary interlocking directorates represent and act as communication channels, which facilitate the flow of important information (Scott 1997) but only to directors who have access to the inner circle and whose ties are strong enough to be seen as a trustworthy member of an interlock.

As elites act in the decentralised information society (Castells 2004) they need to merge in (in)formal networks in order to obtain, maintain and use their privileges and power. Therefore, they merge in coherent networks, called *interlocking directorates*, which are characteristic for the business elite. Interlocking directorates are the result of integration of representatives of boards of directors between companies or corporations, where committees are interlinked and have a common director, who can simultaneously perform this function in two or more corporations and represents the primary decision-making body, and thus a key impact on economic developments (Mizruchi 1992). Within interlocking directorates members of the business elite share the same positions, status, interests and quantity of power, whereby they use interlocking directorates as a source for promotinge the upper-class cohesion, as a crucial prerequisite for their existence (Useem 1984).

Interlocking directorates consist of decision making actors, called the *board of directors*, which represent communication channels and allow privileged individuals access to important information (Mizruchi 1996). In this respect, the merging of directors and executives (i.e. business elite) represents the *inner circle*, as a valuable source of information and guidance on political and economic matters (Useem 1984). Moreover, the inner circle facilitates and incorporate ties and organizational potentials, it allows the flow and access to valuable business information and opportunities, which represent a great advantage for members of these networks in business activities. Members of the business elite are thus guaranteed the sources of crucial information, while other individuals (ordinary citizens) are deprived of the latter.

In addition to social power and influence that the business elite gain with strengthening their social capital via networking, in order to gain relevant information, interlocking directorates facilitate mutual cooperation between managers and executives. The latter are crucial stakeholders with power and possibilities to significantly affect and influence society, its progress and thereby they can stimulate or inhibit society's development.

The emergence and existence of interlocking directorates are crucial for understanding the business elite, its formation, organization and action. Authors who research interlocking directorates (e.g. Mills 1965; Useem 1984; Zajac, 1988; Mizruchi 1992) agree that interlocking directorates imply integration of directors of various boards in different companies. These authors argue about the distribution of interests (inner circle), which allow the elites' groups a disproportionately strong impact, guaranteed by their position in interlocks. Researching interlocking directorates has become a tool of systematic research techniques for analysing publicly available information regarding composition, connections and identities of directors of various companies and organizations (Froud et al. 2006). In this regard, interlocking directorates have become a permanent and important form of networking of the business elite.

Mills (1965) defined interlocks as "*sociological anchor*", which represent a community of interests, as well as unified views and policies that prevail within the class of owners. Useem (1984) focused on the *inner circle* of a bigger number of directors, who establish overall interests of capital and offer a corporate image of long-term needs in business and thus constitute a network of power, which are linked with politics (government, political parties, media, non-profit organizations, etc.). In the past, interlocks have been defined as harmful to the economy, as they allow and encourage clusters, restrict competition and violate the key rule of free market economy. Conversely, contemporary interlocks exist in a modified form, which has become more formal and somehow legalized. The key reasons for merging in contemporary interlocking directorates comprise legitimacy, progress, social cohesion, facilitated flow and access to information, informal communication between directors, control, etc. (Mizruchi 1996). In terms of control, membership in the board of directors is a mitigating circumstance, as it allows the access to internal business information, information asymmetry (Mariolis 1975) and important business transactions. While, the main personal reasons for the establishment of such ties (networks) are social prestige and financial reasons (Zajac 1988), therefore, interlocks work as a tool for promoting the upper class cohesion (Useem 1984), maintained and led by the business elite.

Interlocking directorates operate as closed circles, where membership is reserved for individuals recruited from strategic positions of business organizations. To this point, it is crucial to highlight Granovetter's (1985) research about the functioning of social networks and their impact on businesses and individuals. Granovetter notes that economic actors (enterprises) in relations with each other establish behavioural strategies, structures and operations of networks, characterized by weak or strong ties and relationships. Interpersonal networks allow the most efficient links between the micro and macro level, as networks interactions move from a lower level to a higher one, and again return to the level of small groups. Degree of ties' strength implies dimensions of time, emotional intensity, intimacy and mutual services, which are peculiar to the latter. Weak ties are an essential component of individual chances and individual integration into the community, while strong ties lead to general fragmentation (Granovetter 1983). Ties work as a key adhesive or magnet, which enables integration and existence of these networks in a decentralized society and allow members of the business elite important business opportunities.

Strong ties arise from direct links (networks) between businesses and within enterprise itself, and change into weaker bonds the more their connections are indirect. In practice, this is reflected in the access to information, as these are more accessible to individuals with more powerful (direct) ties (Granovetter 1983). Studies of interlocks in the 70s and early 80s showed that the interlock's ties in large corporations represent the cohesion of the capitalist's class, which enable and facilitate business in effective and dominant political actors, which was reflected in the role of interlocking directorates as active components in various organizations for policy planning (see Mizruchi 1996). Burt (1983) considers such networking as a positive trend, which has an impact on the enterprises' profitability. The latter can be linked with Selznick's (1949) finding that members of the board of directors are individuals, who have the power and influence for improving and eliminating an unstable and unpredictable environment, thereby securing their position in the market. While also individuals who have significant organizational links enhance their reputation. Following Useem's (1984) conceptualisation, the strength of interlocks derives from *political mobilisation of business* since the 70s, as interlocks enabled cooperation between the corporate sector (internal affairs) and politics[4] (external environment).

Besides all this, the essential logic and purpose of interlocks is a special form of insider trading with information and therefore power, nevertheless there is an objective need for their supervision and regulation. Critical views on restrictions of internal trading, focus on arguments that this kind of trading is a form of compensation, which facilitates internal actors (insiders) the access and the flow of confidential information and a rapid implementation and coordination of information with stock prices (Carlton and Fischel 1983). Bhattacharya and Daouk (2002) found that the regulation of internal trade of information causes ambivalent effects such as reduced competition of outside traders, who do not have the access to information. The latter increases their dependence on individuals who

4 As a response to e.g. high tax levels, to too much power vested in labour movements, legislation which hinders enterprise and initiative, etc.

have the possibility to participate and use information. On the other hand, external individuals involved in information trading also deal with information at the expense of uninformed investors (Fishman and Hagerty 1992).

Theories of Bhattacharya and Daouk (2002) or Fishman and Hagerty (1992), etc. can be related to what Moore (1997) stated about reduced access to information, namely it can cause inequalities among individuals. The latter shows the importance of information in the information society, and explains why important individuals (i.e. elite) merge and strengthen their social capital and thus their privileged status and social power. In this context, interlocking directorates act as channels of restrictive information sources, which can facilitate great and important business transactions. This theory also reflects the fact that in many respects, information gained on its importance or even overtook the importance of economic assets. Namely, information has became one of the key determinants and a dominant category that enable benefits, power and influence of individuals who are to be defined as a business elite and thus become key decision makers.

6. Conclusion

The information society brought crucial changes in everyday life as a result of globalisation, among which the most important ones include changes in technology. In fact, when speaking about the information society, we cannot bypass Information and Communisation Technologies (ICT), as a basis for the information society to emerge and operate. The latter works as decentralised unit, wherein the technology infrastructure enables the effectiveness of local and dispersed entities to work and achieve centrally established aims.

An important element of the information society is the so-called network society, which emerged also as a response to the global market, consisting of many dispersed units, which have to operate coherently in order to be successful. Therefore, merging has become an important and indispensable component of the post-modern era and new economy, while networks have become a crucial element for this kind of economy and society to retain. In order to survive and operate successfully in such an economy, business subjects merge in (in)formal networks, which are even more essential for important business transactions, which generate great economic turnouts and allow important businessmen to retain money, power, prestige, etc. Since these businessmen usually occupy strategic positions in important economic organizations, they can be defined as the business elite, who poses economic resources, while also having the access to important business-related information, which enables them to retain their power and prestige or even enhance their status and impact in society.

Despite the fact that the business elite do not represent a cohesive group of individuals, as they pursue individual goals, they have one aim in common, i.e. retention of their privileged status, which is achieved by gathering/merging in networks. Nevertheless, in order to achieve the proper integration in networks and strengthen their ties, the business elite have to gain proper amounts of social capital, which will further enable the acquisition of crucial resources (e.g. money, power, etc.). In this respect, social capital acts as an aggregate of actual and potential resources of networks. The most frequent and common form of the business elites' merging are the so-called *interlocking directorates* as business networks typical for modern capitalism. Interlocking directorates represent interlinked boards of directors as primary decision-making bodies, wherein one person (director) can serve as director of two or more corporations at the same time. Gathering in interlocks facilitates the access to important information, as a crucial product of the information society. Merging in interlocking directorates enables the business elite to access important business-related information and participation at important business transactions. Merging of already powerful and important individuals in (in)formal business networks, which further strengthen their privileged position and power, at the same time represents a prerequisite for business success, as businessmen (business elite) operate in decentralised and global economic sphere. The business elite's merging (networking) further increases its access to information, which is nowadays a valued resource of power and influence, while also in certain cases a much more useful means than economic resources. In fact, information and social capital can maximize the effectiveness of business operations and earnings, thereby further strengthening a privileged (power) position of individuals.

Business elites become the key linkers and thus decision makers in information society, who can successfully influence society's development and progress of all social subsystems (economy, culture, politics). They have succeeded in becoming this strong and important because of their integration and merging in organised and connected networks that operate effectively in a decentralized information society. From this perspective, merging (networking) is crucial for the business elite as it not only enables their existence, but also the increase of social power and influence of these individuals, who can impact and lead the masses. Thus, important businessmen integrate in strong networks wherein strong ties are formed, which represents the so-called inner circle, which ensures their long lasting privileged status. The latter can not be considered as a matter of concern if these ties and networks emerge in the context of a positive collaboration of business elites (i.e. for common good, general development and progress of society, etc.). To this end, networking is an essential element of survival and well functioning of business in the information society, which operates under characteristics of the global market, i.e. many dispersed autonomous economic organizations.

At this point, even though interlocking directorates are legally controlled, we cannot ignore the question of their impact on society and individuals, as the possibility of participation to sources of information arising from inner circles are reserved only for privileged individuals. Proceeding from the fact that information is the crucial source of inequality in the information society, the logical conclusion goes in a direction that limited or denied access to sources of information represents the main source of inequality in the information society.

7. References

Adam, Frane et al.: *The Challenges of Sustained Development: The Role of Socio-Cultural Factors in East-Central Europe*. Central European University Press: Budapest, New York: 2005.

Adler, S. Paul / Kwon, Seok-Woo: "Social Capital: The Good, the Bad, and the Ugly". In: Lesser, L. Eeic (ed.): *Knowledge and Social Capital: Foundations and Applications*. Butterworth-Heinemann: Boston. 2000, pp. 89–115.

Bell, Daniel: *The Coming of Post-Industrial Society: A Venture in Social Forecasting*. Basic Books: New York. 1973.

Bhattacharya, Utpal / Daouk, Hazem: "The world price of insider trading." *Journal of Finance* 57, 2002, pp. 75–108.

Bourdieu, Pierre: The forms of capital. In: Richardson, John G. (ed.): *Handbook of Theory and Research for the Sociology of Education*. Greenwood: New York. 1986, pp. 241–258.

Burt, Ronald S.: *Corporate Profits and Co-optation*. Alliance: New York. 1983.

Carlton, Dennis W. / Fischel, Daniel R.: "The Regulation of Insider Trading". *Stanford Law Review*, 35, 1983, pp. 857–895.

Castells, Manuel: *The Rise of the Network Society. Vol. 1 of The Information Age: Economy, Society and Culture*. Blackwell: Oxford. 1996.

Castells, Manuel: *The Information Age: Economy, Society and Culture, 3 volumes*. Blackwell: Oxford. 1996–1998.

Castells, Manuel: Materials for an exploratory y theory of the network society. *British Journal of Sociology*, 51(1), 2000, pp. 5–24.

Castells, Manuel: The Network Society: A Cross-cultural Perspective. Edward Elgar: Cheltenham, UK et al. 2004.

Dahl, Robert: "The Concept of Power". *Behavioural Science* 22, 1957, pp. 201–15.

Domhoff, G. William: *The Higher Circles*. Random House: New York. 1970.

Etzioni-Halevy, Eva: *The Elite Connection: Problems and Potential of Western Societies* Polity Press: Cambridge. 1993.

Field, G. Lowell / Higley, John / Burton, G. Michael: "A New Elite Framework for Political Sociology". *Cahiers Vilfredo* 28(88), 1990, pp. 149–181.

Figueroa, Adolfo: *Economic elites and social capital*. Research project: International Collaborative Research Project "Confronting the Legacy of Inequality," with the participation of Catholic University of Peru, University of Wisconsin-Madison, University of Natal (South Africa), and IFPRI (Washington DC). 2002.

Figueroa, Adolfo (2008): *Education, labour markets, and inequality in Peru*. Crise Working Paper No. 48. Centre for Research on Inequality, Human Security and Ethnicity. University of Oxford.

Fishman, Michael J. / Hagerty, Kathleen M.: "Insider Trading and the Efficiency of Stock Prices". *The RAND Journal of Economics* 23(1), 1992, pp. 106–122.

Froud, Julie et al.: *Financialization and Strategy. Narrative and Numbers*. Routledge: London. 2006.

Granovetter, Mark: "The strength of weak ties: A network theory revisited". *Sociological Theory* 1, 1983, pp. 201–233.

Granovetter, Mark: "Economic action and social structure: The problem of embeddedness". *American Journal of Sociology* 3(91), 1985, pp. 481–510.

Giles, Martin / Page, Benjamin I.: "Testing Theories of American Politics: Elites, Interest Groups, and Average Citizens". *Perspective on Politics* 12(3), 2014, pp. 564–581.

Gurstein, Michael: "In the context of Community Informatics". In: Sight, Parminder J. / Gurumurthy, Anita / Swamy, Mridula (eds.): *Information Society for the South Series Volume 1: The Political Economy of the Information Society*. IT for Change: Bangole. 2008.

Higley, John: "Elite Theory and Elites". In: Leicht, Kevin T. / Jenkins, J. Craig (eds.): *Handbook of Politics: State and Society in Global Perspective*. 2010, pp. 161–176.

Keller, Suzanne: *Beyond the Ruling Class: Strategic Elites in Modern Society*. Transaction Publishers: New Brunswick and London. 1991.

Mariolis, Peter: "Interlocking Directorates and Control of Corporations: The Theory of Bank Control". *Social Science Quarterly* 56, 1975, pp. 425–39.

Mills, Wright, C. (1965): *Elita oblasti*. Državna založba Slovenije, Ljubljana.

Mizruchi, Mark, S.: *The Structure of Corporate Political Action*. Harvard University Press: Cambridge. 1992.

Mizruchi, Mark, S.: "What Do Interlocks Do? An Analysis, Critique, and Assessment of Research on Interlocking Directorates". *Annual Review of Sociology* 22, 1996, pp. 271–298.

Mosca, Gaetano: *The Ruling Class*. McGraw: New York. 1939.

Moore, Nick: "The information society". In: Courrier, Yves (ed.): *World information*. UNESCO Publishing: Paris. 1997, pp. 271–284.

Pareto, Vilfredo: *The Mind and Society.* Harcourt Brace: New York. 1935.

Parsons, Talcott: *Structure and Process in Modern Societies*. The Free Press: New York. 1960.

Portes, Alejandro: "Social Capital: Its Origins and Applications in Contemporary Sociology". *Annual Review of Sociology* 24, 1998, pp. 1–24.

Prijon, Lea: "Abuse of power and influence of elites through informal networks". In: Pandiloska Jurak, Alenka / Pinterič, Uroš (eds.) *Contemporary world between freedom and security*. Vega: Ljubljana. 2010, pp. 377–402.

Prijon, Lea: "Merging economic and political elite: Slovenian interlocks". *Raziskave in razprave* 8(1), 2015a, pp. 210–258.

Prijon Lea: *Poslovna elita: Analiza pogojev za njeno formiranje in delovanje*. Vega: Ljubljana. 2015b.

Prijon, Egon / Prijon, Lea: "The impact of exogenous factors on business elite's formation: Comparing societies". *Revija za znanstveno odličnost* 4(2), 2015 pp. A12-A36.

Scott, John. *Corporate Business and Capitalist Classes*. Oxford University Press: Oxford. 1997.

Selznick, Philip: *TVA and the Grass Roots*. Harper & Row: New York. 1949.

Szalacha, J.: "Interlocking directorates and possible conflict of interests". *Polish Sociological Review* 2(174), 2011, pp. 205–223.

The Clayton Antitrust Act. Pub.L. 63–212, 38, October 15, 1914, Stat. 730.

The Sherman Antitrust Act. Pub. L. 94–435, 90, July 2, 1890, Stat. 1383.

Tomšič, Matevž: *Elites in Post-Communist Societies*. Vega Press, Ltd: London. 2011.

Useem, Michael: *Inner Circle: Large Corporations and the Rise of Business Political Activity in the U.S. and U.K.* Oxford University Press: New York, Oxford. 1984.

Webster, Frank: *Theories of the Information Society*. Third edition. Routledge: London and New York. 2006.

Webster, Frank: *Theories of the Information Society*. Fourth edition. Routledge: London and New York. 2014.

Zajac, Edward J.: "Interlocking Directorates as an Interorganizational Strategy". *Acad. Manage* 31, 1988 pp. 428–38.

Jana Suklan
School of Advanced Social Studies in Nova Gorica
Gregorčičeva 19, 5000 Nova Gorica, Slovenia
jana.suklan@fuds.si

Vesna Žabkar
Faculty of Economics, University of Ljubljana
Kardeljeva ploščad 17, 1000 Ljubljana, Slovenia
vesna.zabkar@ef.uni-lj.si

How to Benefit from Cross-Media Synergies in the Information Society

Abstract: *In the information society the exchange of information is the predominant economic activity. The communication media introduces us to the mass media. Mass media channels of communication represent the predominant share of our marketing environment. The marketing environment is characterized by multichannel shopping linked with more choice in advertising channels. Analysing the data using specific market related knowledge increases the understanding of patterns in consumer reactions to advertising. In their advertising campaign, marketers can choose between different communication channels to persuade customers to buy services or products from their company. When the presence of synergy is established optimal media budget allocation, the optimal media mix and advertising carry-over effects differ. The purpose of the present paper is to present a synergy model that provides an extension of a marketing-mix approach with insights into how different components interact over time. We applied a statistical modelling technique primarily used in financial markets to model advertising effectiveness.*

Keywords: *cross-media synergy, multichannel shopping, advertising channels, advertising models.*

1. Introduction

Advertising is a practice of getting people's attention to one's product, service or idea through paid messages in the media. Thus it is important that messages are delivered and received executional and strategically consistent so that they can create coherent perceptions among customers and other stakeholders (Duncan and Moriarty 1998 p. 45). Therefore the integration of the various marketing communication's messages is required (Mehta 2000 p. 69). The multichannel marketing environment presents a major marketing investment for companies and is as such a challenge for marketers. However the mass media and advanced information technology can helping us grasp the knowledge base, skills, and perspective that cut across the industries (LaRose 1995 p. 315). Analysing the data using specific market related knowledge increases the understanding of patterns in consumer reactions to advertising. When we prove that the presence of synergy between communication channels exists, we should advise brand managers to determine the total media budget and media mix differently (Naik and Raman 2003, pp. 380–381; Raman and Naik 2006, p. 442). Changes in advertising lead marketers to use an integrated marketing communications (IMC) perspective in advertising and promotion. IMC calls for planning and coordination of various communication channels while doing marketing and promotion programs (Kitchen et al. 2004, pp. 21–25). The number of companies that acquire an IMC perspective for advertising and promotion is rising. Within such a perspective, marketers are coordinating elements of the promotional mix and other marketing activities to communicate with customers. The overall marketing process, consumer behaviour, and communications theory are the prerequisites for effective planning, implementing and evaluation of IMC programs (Belch and Belch 2003). The challenge for marketers is to choose among different communication options the right ones for a communication campaign in order to maximize "push" and "pull" effects in the multimedia environment (Keller 2010, p. 59) as well as long-term, profitable relationships with the potential customers. Companies engaging in multichannel communication should search for an optimal balance in the frequency of advertising, an adequate communication channels selection and alignment of channels with customers' preferences. Studies indicated that by exceeding the optimal level of communication negative reactions or avoidance could be expected. When applying an IMC strategy in the company, marketers should take into consideration both the cannibalisation of one channel over the other or wear out signs due to high frequency of the advertising repletion which drive customers away from a company (Schmidt and Eisend 2015, p. 13). With a balance of behavioural and critical perspectives, communications media encourage us to reconsider the effects and implications of the traditional and new media, preparing us to thrive in a new communications environment in the information society (Straubhaar et al. 2013, p. 30).

The structure of the paper focuses firstly on a literature review in order to draw a theoretical outline and to summarize findings of previous researches. Secondly we describe the vector autoregression method that we will apply to our models and why we believe applying a method used mainly to forecast macroeconomic variables could also be applied on sales data. To justify the method used we checked the diagnostics and discussed empirical results. According to the given results we prepared implications for the marketers on how to respond to allocate funds in the presence of synergy.

2. Literature Review

The marketing environment is characterized by multichannel shopping linked with more choice in advertising channels. In their advertising campaign, marketers can choose between different communication channels to persuade customers to buy services or products from their company. The usage of multimedia in one campaign is widespread. According to the study by Chang and Thorson (2004) marketers used multiple communication tools or channels within a single campaign to broaden the reach of a campaign to get to different target groups and to prepare a fruitful ground for synergy to sprout and thus create the greatest persuasion effect (Dijkstra et al. 2005, p. 379). Using multiple communication channels to communicate with a specific segment of the target audience generates extra "synergetic" effects and can improve campaign results (Chang and Thorson 2004, pp. 76–79). As marketers are combining different advertising and promotional tools we can talk of the IMC concept, new advertising, orchestration, and seamless communication (Belch and Belch 2003; Kitchen et al. 2004, p. 22; Christensen et al. 2005, p. 157; Groom and Frei 2008, pp. 7–15). Much before we perceive synergy between inter-correlated and in aligned communication channels the shift should happen on another level, inside the organisation. Once managers prepare a fruitful ground for synergy to sprout then the maximized measurable results can be achieved. Units in the company generate greater value when working together, because there is shared know-how, shared tangible resources, pooled negotiating power, coordinated strategies, vertical integration (reduced costs) and combined business creation (internal joint ventures) (Goold and Campbell 1998, pp. 133–136). A broad definition of IMC states that it "*is a concept of marketing communication planning that recognises the added value of using a comprehensive plan to evaluate the strategic roles of a variety of communication disciplines. It combines these disciplines to provide clarity, consistency and maximum communication impact*" (Schultz et al. 1993, p. 93). The definition could be extended in a way that defines the IMC as a strategic control process and persuade customers through all messages and setting up a feedback mechanism in order to create and maintain profitable relationships with customers and other stakeholders (Duncan and Caywood 1996). Strategic integration of various communication functions makes the IMC a new approach able to reach consumers and other stakeholders (Duncan and Everett 1993, p. 30; Kitchen et al. 2004, 17–18). It integrates all of the companies' marketing activities by coordinating their advertising efforts. An IMC is used to maximize the return on investment in marketing and promotion since it avoids duplication, creates synergy, develops coordinated marketing communications programs (Belch and Belch 2003, p. 11). Dekimpe and Hanssens (1999) found that in many cases funds are directed to the communication channel that have the greatest long-term benefit. Synergy between communication channels is defined as a combined effect of multiple activities when they exceed the sum of their individual effects (Belch and Belch 1998, p. 11). Empirically synergy is produced as an interaction term between mixed media communications targeted at the same market segments aiming to increase the effectiveness of each media due to the presence of the other (Prasad and Sethi 2009, p. 604). Synergy emerges when companies employ multiple communication channels in a single marketing campaign. Similarity and congruence of the ad execution across the different media is assumed to be an important factor in creating cross-media synergy. In order to make consumers realize that ads are part of one integrated campaign, integration should link different ads together (Voorveld and Valkenburg 2013, 189–195).The purpose of an IMC is to manage all marketing activities that impact sales, profits, and brand equity simultaneously. The IMC model emphasizes the role of the joint effects or synergies generated due to the synchronization of multiple activities (Raman and Naik 2006, p. 437). In the interaction between two or more media one is likely to be dominant either in the aspect of costs or revenue. The negative synergy can also raise when the consumer's exposure to one media occurs at the expense of the other (Assael 2011, pp. 43–50) or when one media is taken as a substitute for another (Kollmann et al. 2012, pp. 187–190). The cannibalisation effect or negative synergy means that too much advertising on one channel can lead to losses in the others.

3. Empirical Analyses

Throughout the analysis we aimed to understand the relationships and effects of the relevant indicators for the model. The purpose of the paper is to check for the presence of media synergy between communication channels and to empirically validate the effectiveness of the regression model. From the aggregated market data we are able to track sales separately for each communication channel. The method applied in the empirical analysis is primarily used in financial markets to forecast macroeconomic variables (Bernanke et al. 2005, p. 389; Pauwels et al. 2004, pp. 598–600; Wang et al. 2013, p. 1225). According to Giannone et al. (2015) vector autoregressions (VARs) are flexible time series models that can capture complex dynamic interrelationships among variables, they perform well both out-of-sample forecasting as well as factor models.

3.1 Data Description

Investment in advertising and sales data of a retail service were analysed empirically in order to test the conceptual model of the IMC principle. The analysis refers to an established brand of high involvement household product. Investment in advertising and promotion set for the brand comprises weekly time series, starting from year 2007 till the year 2012. The sales data consist of television, on-line, wholesale (reseller) and retail (company stores) sales measured in value of units sold. Proprietary data consists of the number of commercial minutes (for advertising on air) on national public and commercial television stations in the national market, while online activities are expressed in number of online visitors. Visits represent the number of individual sessions initiated by all the visitors to the site. Data collection is detailed; each channel has its own phone or track number, which allows us to track sales separately for each channel. We constructed dummy variables to capture seasonal effects of sales; we identified drops (in March and August) and peaks (in December). In order to maintain confidentiality, we are not in a position to reveal any of proprietary data, including the identity of the company or its basic products. All the data were previously rescaled.

The company appeal to the consumers with the use of coordinated messages delivered to them through integrated communication channels. In terms of business process modelling, it is appropriate to develop a complex model of interactions. Our model could be divided in two parts. Input data includes advertising and promotion on the one hand, and the sales on the other. Sales may be described as a function of the all independent variables. In the model we include both direct and indirect sales channels, and personal and mass communication channels. In terms of channels of distribution, we included online as a direct and interactive channel, then resellers (supermarket, catalogue showroom) as an indirect channel. In terms of marketing communications, personal communications were employed through personal selling (company stores) and mass communications through advertising on television (Keller 2010, p. 59).

3.2 Model setting and definition

Advertising effectiveness is typically measured through econometric models that measure the impact of varying levels of advertising on sales. Here current effect of advertising is defined as the change in sales caused by an exposure of advertising occurring at the same time period as the exposure (Tellis 2006, p. 506). The carryover effect of advertising is a portion of its effect that occurs in time periods following the advertising.

To estimate time-series regression parameters a large number of methods have been developed. When choosing between different advertising models the advertising-sales relationship must be specified. Apart from that we classify models according to the structure of the system, operationalization of the variables and specification of the random error (Rao 1972, pp. 180–181). Since time-series data violate the assumptions that the error terms corresponding to different points in time that are not correlated (error terms are serially or autocorrelated), thus the least squares (OLS) analysis produces biased estimates of the standard errors of the regression coefficients (Choudhury et al. 1999, pp. 342–343). The vector autoregression (VAR) models are a sort of econometric model designed to deal with linear interdependencies between multiple time series (Tong et al. 2011, p. 301). VAR models treat all variables included in it symmetrically; each variable has an equation explaining its evolution based on its own lags and the lags of the other model variables (Zeng et al. 2014, p. 2). Vector autoregression (VAR) model is one of the most successful, flexible, and easy to use models for the analysis of multivariate time series. It is a natural extension of the univariate autoregressive model to dynamic multivariate time series (Zivot and Wang 2006, pp. 384–395). In the case of reduction VAR every variable is explained as a linear function of its own past values, past values of all the other variables in the model and serially correlated errors (Stock and Watson 2001, pp. 112–113).

According to the Box and Jenkins procedure we first removed systematic variation in advertising and eliminate trends and seasonality in sales (Helmer and Johansson 1977, pp. 227–239), and then we transformed the time series with the use of natural logarithm and differentiate. The aim of transformation such as differencing, log-differencing and detrending is to reduce non-stationarity of the time series. Time series is said to be stationary if there is no systematic change in mean (no trend), if there is no systematic change in variance and if strictly periodic variations have been removed (Chatfield 2013). We checked for stationary using Augmented Dickey-Fuller t-statistic (see Table 1).

Table 1: Augmented Dickey-Fuller Test

Variable	Augmented Dickey-Fuller Test			
Data in % change	F statistic	Estimate	Test statistic: tau	p-value
Televending sales rate	-0,079	-0,74	-11,0	0,000
Online sales rate	-0,429	-0,28	-5,66	0,000
Retail sales rate	-0,417	-0,28	-5,76	0,000
Resellers sales rate	0,038	-0,08	-2,84	0,004

Data analysis was performed using Gretl statistical software (Baiocchi and Distaso 2003). VAR analysis included four endogenous sales variables, three exogenous input variables tracking advertising effort synergy and two seasonal dummy variables. In the next step residual autocorrelation, residuals normality and residual heteroskedasticity are applied to the estimated model (see Table 2).

Table 2: Vector autoregression

Equation	S.E. of regression	Adjusted R-squared	P-value (F)	Durbin-Watson
Televending sales rate	0,435479	0,258177	$7{,}47*10^{-7}$	2,152625
Online sales rate	0,230391	0,286623	$1{,}01*10^{-11}$	2,125170
Retail sales rate	0,019009	0,333170	$5{,}65*10^{-15}$	2,031185
Resellers sales rate	0,359172	0,643409	$3{,}2410^{-80}$	1,955937

Log-likelihood = 374,09; Determinant of covariance matrix = $2{,}95*10^{-7}$; AIC = -2,8580; BIC = -1,4870; HQC = -2,3034; Portmanteau test: LB(48) = 739,844, df = 720 [0,2961]; Likelihood ratio test: Chi-square(16) = 54,1998 [0,0000].

Residual autocorrelation can be tested with the help of Ljung-Box Q'statistics, where the null hypothesis denotes the absence of autocorrelation. From Table 3 we can see Ljung-Box Q' is statistically significant at the 0,05 rate for most variables.

Table 3: Residual autocorrelation

Equation	Residual Autocorrelations		
	Ljung-Box Q'	df	p-value
Televending sales rate	31,91	14	0,004
Online sales rate	13,84	14	0,462
Retail sales rate	24,05	14	0,045
Resellers sales rate	26,25	14	0,024

Residual autocorrelation is still present in the equation where the online sales rate plays the role of the dependent variable. The residual distribution normality will be tested by the use of the Jarque-Bera test. Residual normality tests show that the residuals of the models are not normally distributed meaning the departure from the theoretical distribution (see Table 4).

Table 4: Residual normality test

Equation	Residual Normality Test	
	Jarque-Bera test	p-value
Televending sales rate	17,43	0,000
Online sales rate	10,12	0,006
Retail sales rate	12,57	0,002
Resellers sales rate	252,1	0,000

Heteroskedasticity of the residuals will be tested using Breusch-Pagan statistics. With Breusch-Pagan statistics we are testing the null hypothesis that residuals are homoscedastic meaning that the variance is constant in time. The third model provides homoscedastic residuals, thus produce nonbiased estimates (see Table 5).

Table 5: Residual Heteroskedasticity

Equation	Residual Heteroskedasticity		
	Test statistic: LM	df	p-value
Televending sales rate	45,37	14	0,0000
Online sales rate	41,91	14	0,0001
Retail sales rate	16,53	14	0,2822

In the other cases heteroscedasticity is present implying that we need to interpret robust White's heteroscedasticity - consistent estimate (see Table 2). Visitors on web page contribute significantly to televending sales rate and online sales rate, while lagged synergetic effect of minutes of input in television advertising and online visitors have a significant effect on televending, online and reseller's sales rate. In the first case synergy is negative, indicating a cannibalisation effect of one communication channel over the other. Alternating positive and negative lagged synergy affects also online sales rate, while positive synergy is focused towards reseller's sales rate (Table 6).

Table 6: VAR Models Estimations Including Lags

Equation 1: TELEVENDING	Coefficient	Std. Error	t-ratio	p-value	
const	−2,28035	1,07102	−2,1291	0,0346	**
TELEVENDING _1	−0,526038	0,0865999	−6,0743	<0,0001	***
TELEVENDING _2	−0,384022	0,119745	−3,2070	0,0016	***
TELEVENDING _3	−0,282191	0,0876084	−3,2210	0,0015	***
ONLINE SALES _1	−0,156975	0,15435	−1,0170	0,3105	
ONLINE SALES _2	−0,18546	0,188423	−0,9843	0,3263	
ONLINE SALES _3	−0,01193	0,134151	−0,0889	0,9292	
RETAIL SALES_1	−1,55838	1,34319	−1,1602	0,2475	
RETAIL SALES _2	0,333204	1,67086	0,1994	0,8422	
RETAIL SALES _3	0,87881	1,37681	0,6383	0,5241	
RESELLERS _1	−0,0489511	0,0900711	−0,5435	0,5875	
RESELLERS _2	0,0215378	0,117917	0,1827	0,8553	
RESELLERS _3	0,0527982	0,0875286	0,6032	0,5471	
MINUTES ON TELEVISION	0,00043388	0,104046	0,0042	0,9967	
VISITORS ON WEBPAGE	0,28035	0,12038	2,3289	0,0210	**
SYNERGY	−0,375761	1,22167	−0,3076	0,7588	
SYNERGY _1	−0,783269	1,20697	−0,6490	0,5172	
SYNERGY _2	−0,828465	1,24722	−0,6642	0,5074	
SYNERGY _3	−3,30025	1,13877	−2,8981	0,0042	***
SEASONAL PEAK	−0,0659697	0,0902364	−0,7311	0,4657	
SEASONAL DROP	−0,1008	0,0754545	−1,3359	0,1832	
Equation 2: ONLINE SALES	**Coefficient**	**Std. Error**	**t-ratio**	**p-value**	
const	−0,835855	0,485881	−1,7203	0,0871	*
TELEVENDING _1	0,126003	0,0413345	3,0484	0,0026	***
TELEVENDING _2	0,108899	0,0444015	2,4526	0,0151	**
TELEVENDING _3	0,0678382	0,035217	1,9263	0,0556	*
ONLINE SALES _1	−0,568242	0,0860252	−6,6055	<0,0001	***
ONLINE SALES _2	−0,219414	0,0744222	−2,9482	0,0036	***
ONLINE SALES _3	−0,071672	0,0667881	−1,0731	0,2846	
RETAIL SALES_1	−0,139824	0,739802	−0,1890	0,8503	
RETAIL SALES _2	−0,0403638	0,798424	−0,0506	0,9597	
RETAIL SALES _3	−0,673998	0,808321	−0,8338	0,4055	
RESELLERS _1	0,0538301	0,0351211	1,5327	0,1271	
RESELLERS _2	−0,0632121	0,0432472	−1,4616	0,1456	
RESELLERS _3	0,0270241	0,040177	0,6726	0,5020	
MINUTES ON TELEVISION	−0,0407695	0,0410115	−0,9941	0,3215	

Equation 2: ONLINE SALES	Coefficient	Std. Error	t-ratio	p-value	
VISITORS ON WEBPAGE	0,127782	0,0596343	2,1428	0,0335	**
SYNERGY	−1,10161	0,655296	−1,6811	0,0945	*
SYNERGY _1	−0,723776	0,578603	−1,2509	0,2126	
SYNERGY _2	1,01041	0,610794	1,6543	0,0998	*
SYNERGY _3	−0,966218	0,466261	−2,0723	0,0397	**
SEASONAL PEAK	0,0452468	0,0518329	0,8729	0,3838	
SEASONAL DROP	0,0190349	0,0386241	0,4928	0,6227	
Equation 3: RETAIL SALES	Coefficient	Std. Error	t-ratio	p-value	
const	−0,0224628	0,0431133	−0,5210	0,6030	
TELEVENDING _1	0,000689365	0,00253427	0,2720	0,7859	
TELEVENDING _2	−0,00262711	0,00270319	−0,9719	0,3324	
TELEVENDING _3	0,00140654	0,00235244	0,5979	0,5506	
ONLINE SALES_1	0,00860677	0,00545758	1,5770	0,1165	
ONLINE SALES_2	0,0275771	0,00668633	4,1244	<0,0001	***
ONLINE SALES_3	0,0153608	0,00420781	3,6505	0,0003	***
RETAIL SALES _1	−0,699415	0,0690092	−10,1351	<0,0001	***
RETAIL SALES _2	−0,443129	0,0815963	−5,4307	<0,0001	***
RETAIL SALES _3	−0,222691	0,0791211	−2,8146	0,0054	***
RESELLERS _1	0,00162339	0,00330276	0,4915	0,6236	
RESELLERS _2	−0,00712962	0,00409288	−1,7420	0,0832	*
RESELLERS _3	0,00326002	0,00318639	1,0231	0,3076	
MINUTES ON TELEVISION	−0,00174365	0,0033727	−0,5170	0,6058	
VISITORS ON WEBPAGE	0,00404313	0,00500033	0,8086	0,4198	
SYNERGY	−0,0295348	0,0456031	−0,6476	0,5180	
SYNERGY _1	0,0514066	0,04118	1,2483	0,2135	
SYNERGY _2	−0,00262056	0,0456391	−0,0574	0,9543	
SYNERGY _3	0,0473267	0,0414976	1,1405	0,2556	
SEASONAL PEAK	0,00770004	0,0037084	2,0764	0,0393	**
SEASONAL DROP	−0,00148607	0,0031627	−0,4699	0,6390	
Equation 4: Resellers	Coefficient	Std. Error	t-ratio	p-value	
const	−0,213474	0,74506	−0,2865	0,7748	
TELEVENDING _1	0,0262709	0,0603219	0,4355	0,6637	
TELEVENDING _2	0,0377565	0,0886137	0,4261	0,6706	
TELEVENDING _3	0,00522909	0,0492586	0,1062	0,9156	
ONLINE SALES _1	−0,0796447	0,114188	−0,6975	0,4864	
ONLINE SALES _2	−0,125187	0,128647	−0,9731	0,3318	
ONLINE SALES _3	0,121707	0,092291	1,3187	0,1889	

Equation 4: Resellers	Coefficient	Std. Error	t-ratio	p-value	
RETAIL SALES _1	−1,76898	1,27158	−1,3912	0,1659	
RETAIL SALES _2	−1,93606	1,15717	−1,6731	0,0960	*
RETAIL SALES _3	−2,99924	1,0776	−2,7833	0,0059	***
RESELLERS_1	0,807683	0,0529559	15,2520	<0,0001	***
RESELLERS _2	−0,00813616	0,037793	−0,2153	0,8298	
RESELLERS _3	−0,172467	0,0477585	−3,6112	0,0004	***
MINUTES ON TELEVISION	0,0351319	0,0659006	0,5331	0,5946	
VISITORS ON WEBPAGE	0,00199722	0,0802766	0,0249	0,9802	
SYNERGY	0,370292	0,696608	0,5316	0,5957	
SYNERGY _1	0,967297	0,610857	1,5835	0,1150	
SYNERGY _2	1,76331	1,01068	1,7447	0,0827	*
SYNERGY _3	0,888486	0,87534	1,0150	0,3114	
SEASONAL PEAK	0,261893	0,0805608	3,2509	0,0014	***
SEASONAL DROP	−0,0853887	0,0714971	−1,1943	0,2339	

Legend: _1, _2, _3…lagged sales.

4. Conclusions and Implications

Vector autoregression estimation computed provided the structure of intercorrelations between endogenous and exogenous variables included in the model. From the results we could identify an effect of televending sales rate on online sales rate, then an effect of lagged online sales rate on retail sales rate and a two sided effect between retail sales rate and reseller sales rate.

In all models, past sales affect the actual sales significantly. Lagged sales of televending sales rate is negative, implying that the most the customers already bought the less they will need now or in the future. Dealing with a durable household good this is very plausible. The same is true for online sales and retail sales. The effect of the past on present resellers sales rate is alternating from positive to negative. From the managerial point of view the existence and the direction of synergy are important. As already noted, when synergy is detected the allocation of funds should be done differently (Naik and Raman 2003, p. 377). Then budget allocation must be redefined according to circumstances. Budget allocation based on synergy is influenced by the dominance of one media over the other (Assael 2011, p. 45). Sometimes positive synergy is demonstrated, however, also negative synergy provides even more valuable information for the marketer. Contrary to circumstances when positive synergy is present, when cannibalisation is revealed the former allocation strategy must be revised. In the case of televending and online sales rate, cannibalisation is present between input in television and the web. When negative synergy is present then it is important for marketers to spot the source of cannibalisation. Whether one communication channel cannibalizes sales from the other, action should be taken in either omitting the ineffective channel from the marketing strategy or reallocating fewer funds to the aggressive channel in order not to affect total sales negatively. When we omit the ineffective channel from the marketing strategy we can hope that the remaining channels will cooperate resulting in positive synergy on sales. In our case however both communication channels are important for total sales. Thus the elimination of the less effective is not an option. When dealing with cannibalisation among key communication channels reallocation is the only option. Thus we allocate more funds to the more effective communication channel, but when the channel cannibalizes the others, then the input in the channel is lower than in the absence of negative synergetic effect in the model. Total funds designed for advertising should drop. Analytical tools can help to evaluate the response of advertising on sales. The overall contribution of advertising is the key element on the basis of which we make a conclusion if investment in a cannibalistic communication channel could be justified. If so, the investment in such a channel should diminish. The question for the marketers remains whether the benefits brought about by investing in an effective channel while cannibalizing others, outweigh the potential loss in terms of sales if fewer funds are allocated in such a channel.

Since media technology changes with every generation, traditional mass media channels are combining with new madia in ways that change our media consumption patterns, as well as our lives, and the societies in a holistic perspective (Straubhaar et al., 2013, p. 30).

5. References

Assael, Henry: „From silos to synergy.“ *Journal of Advertising Research* 51(1), 2011, pp. 42–58.

Baiocchi, Giovanni / Walter Distaso: "GRETL: Econometric software for the GNU generation." *Journal of Applied Econometrics* 18(1), 2003, pp. 105–110.

Belch, George E. / Michael Belch A.: *Advertising and promotion: An integrated marketing communications perspective.* The McGraw– Hill, 2003.

Bernanke, Ben S. et al.: "Measuring the effects of monetary policy: a factor-augmented vector autoregressive (FAVAR) approach". *National Bureau of Economic Research*, No. w10220, 2004.

Christensen, Thøger et al.: "Integrated marketing communication and postmodernity: an odd couple?" *Corporate Communications: An International Journal* 10(2), 2005, pp. 156–167.

Duncan, Tom / Sandra Moriarty E.: "A communication-based marketing model for managing relationships." *The Journal of marketing*, 1998, pp. 1–13.

Duncan, Thomas R. / Stephen Everett E.: "Client perceptions of integrated marketing communications." *Journal of Advertising Research* 33(3), 1993, pp. 30–40.

Giannone, Domenico et al.: "Prior selection for vector autoregressions." *Review of Economics and Statistics* 97(2), 2015, pp. 436–451.

Goold, Michael / Andrew Campbell: "Desperately seeking synergy." *Harvard Business Review* 76(5), 1998, pp. 131–143.

Groom, Alyssa S.: "Integrated Marketing Communication Anticipating the" Age of Engage""." *Communication research trends* 27(4), 2008, pp. 3–19.

Helmer, Richard M. / Johny Johansson K.: "An exposition of the Box-Jenkins transfer function analysis with an application to the advertising-sales relationship." *Journal of Marketing Research* 1977, pp. 227–239.

Keller, Kevin L.: "Brand equity management in a multichannel, multimedia retail environment." *Journal of Interactive Marketing* 24(2), 2010, pp. 58–70.

Kitchen, Philip J., et al.: "The emergence of IMC: a theoretical perspective." *Journal of advertising research* 44(1), 2004, pp. 19–30.

Kitchen, Philip J., et al.: "Will agencies ever "get" (or understand) IMC?." *European journal of marketing* 38(11/12), 2004, pp. 1417–1436.

LaRose, Robert: *Communications media in the information society*. Wadsworth Publ. Co., 1995.

Mehta, Abhilasha: "Advertising attitudes and advertising effectiveness." *Journal of advertising research* 40(3), 2000, pp. 67–72.

Naik, Prasad A. / Kalyan Raman: "Understanding the impact of synergy in multimedia communications." *Journal of Marketing Research* 40(4), 2003, pp. 375–388.

Pauwels, Koen, et al.: "Modeling marketing dynamics by time series econometrics." *Marketing Letters* 15(4), 2004, pp. 167–183.

Raman, Kalyan / Prasad Naik A.: *Integrated marketing communications in retailing*. (Retailing in the 21st Century). Springer Berlin Heidelberg, 2006.

Rao, Vithala R.: "Alternative econometric models of sales-advertising relationships." *Journal of Marketing Research* 1972, pp. 177–181.

Schmidt, Susanne / Martin Eisend: "Advertising Repetition: A Meta-Analysis on Effective Frequency in Advertising." *Journal of Advertising* 44(4), 2015, pp. 415–428.

Straubhaar, Joseph et al.: *Media now: Understanding media, culture, and technology.* Cengage Learning, 2013.

Tellis, Gerard J.: "Modeling marketing mix." Handbook of marketing research (2006): 506–522.

Tong, Hengqing, T. et al.: "Simultaneous Equations Models and Distributed Lag Models." *Developing Econometrics*, pp. 215–251.

Wang, Yudong et al.: "Oil price shocks and stock market activities: Evidence from oil-importing and oil-exporting countries." *Journal of Comparative Economics* 41(4), 2013, pp. 1220–1239.

Zeng, Tao et al.: "Deviance Information Criterion for Comparing VAR Models." *Advances in Econometrics: Essays in Honor of Peter CB Phillips* 2014, pp. 615–637.

Zivot, Eric / Jiahui Wang: "Vector autoregressive models for multivariate time series." *Modeling Financial Time Series with S-PLUS* 2006, pp. 385–429.

Politics: Old Challenges, New Perspectives

Matevž Tomšič and Petra Kleindienst
School of Advanced Social Studies in Nova Gorica
Gregorčičeva 19, 5000 Nova Gorica, Slovenia
matevz.tomsic, petra.kleindienst@fuds.si

E-Participation, E-democracy and Political Engagement of the Citizenry in Central and Eastern Europe

Abstract: *In many countries, one can perceive the long-term decline in political trust in combination with a decrease of electoral turnout. Thereby mechanisms for enhancing political participation and upgrading of political competence of the citizens in contemporary democracies could be very substantial. The paper explores e-participative tools in Central and Eastern Europe, based on utilisation of information-communication technology, aiming to stimulate political and civil engagement of the citizenry. The innovative forms of political activity in the terms of e-participation could increase political competence of the citizenry, enabling them to make more informed political choices. This is the topic of our paper where we discuss practices and possibilities for introduction of innovative tools of e-participation on the example of EUVox platform.*

Keywords: *participation, democracy, information-communication technology, EUVox, Central and Eastern Europe*

1. Introduction

The political space in modern democracies went through a profound change in the last couple of decades. It has to cope with the challenge of decreasing political participation and the lack of confidence in political institutions and elected candidates (Putnam et al. 2002; Dalton 2004). This can be related to a variety of factors that include higher education levels, post-modern scepticism towards any authority (Inglehart 1997) and increasing individualisation (Beck and Beck-Gernsheim 2002). Some traditional forms of political and civil participation like involvement in political parties and different civil society organisations have been losing this importance (what resulted in their decreased membership). In many countries, the long-term decline in political trust is coupled with a decrease of electoral turnout.

These developments are even more evident in the former communist countries from Central and Eastern Europe (CEE) that only two and a half decades ago launched their 'democratic transformation'. They are characterised by both mistrust in political institutions and decreasing political participation, as manifested in levels of voting turnout that is significantly lower than in 'old' democracies (Makarovič and Tomšič 2010). Lower confidence in political institutions is not necessary problematic for democratic stability. It could mean dissatisfaction of democratic performance and not withdrawal of democratic support, thus providing an impetus for necessary political reforms (Dalton 2002, 2004; Tarrow 2002). The lack of confidence in political institutions, in combination with apathy, could increase transaction costs and harm the performance of (already relatively weak) institutions. Furthermore, a vicious circle appears where decreasing trust results in political and civic apathy which further harms the level of political trust.

In this regard, the new forms of political activity in the terms of e-participation, based on utilisation of information-communication technology, could stimulate political and civil engagement of the citizenry. According to the objectives of democratic functioning there is a broad range of online tools trying to foster transparent and inclusive democratic participation. Utilisation of these tools can increase the level of trust in key political institutions, thus breaking the above-mentioned vicious circle. Furthermore, certain e-participative tools are designed to increase political competence of the citizenry, enabling them to make more informed political choices. In our chapter, we will discuss practices and possibilities for e-participation in former communist countries.

2. Participation as one of the main components of contemporary democracy

Political participation is one on the main components of modern polyarchy as the type of political setting that is in reality the closest to democratic ideals (Dahl 1971, 1989). It can be perceived as a component of inclusiveness of modern democratic systems (Tomšič and Vehovar 2012). It refers to the involvement of the people in decision-making processes through more or less formalised channels. It consists of all activities that are in one way or another intended to affect actions and results of political system (Goel in Smith et al. 1980, p. 76). It is related to positive freedom as perceived by Berlin (1969), i.e. ability of the citizenry to 'rule themselves' on the basis of their public engagement.

The most elementary mechanism of a political participation is taking part in elections at different levels of polity where candidates compete for the most important political positions. In this context, parliamentary and presidential elections play the most important role. We are speaking about participation in terms of both involvement in the process of selection of candidates (voting) as well as running for office (being elected). Nowadays, inclusiveness of voting rights – in terms of inclusion of all capable adults in the electorate – is one of the key elements of democratic life. Namely, in the case when a certain social group is banned from participation in electoral processes, its interests cannot be properly represented in the political realm (Dahl 1998, p. 77). This leads either to their complete social exclusion or to paternalist guardianship of ruling circles over what keeps them in a position of dependency and inability to determine their own destiny.

Although it is clear that electoral participation is an important attribute of the democracy policy, it is a question of how 'high' it should be. Namely, some established democracies (especially the US) are in fact known for a relatively low turnout while in some others, there is traditionally high voter turnout. It should be noted that in some democratic countries, voting (at least at the national level) is mandatory, that certainly affects the (higher) level of voter turnout. The frequency of elections is also different since in countries with federal rule they are usually more often what affect the (lower) average participation. In addition, the motives of citizens to vote are different, meaning that it is not just about a personal desire to participate in politics. Often some other, more traditional motives like sense of duty, etc., may also play an important role in this respect.

In this regard, it is probably true that in societies with a stable democracy, (very) high voter turnout is not necessarily a sign of a democratic deficiency, although it can become problematic in the case when the electoral abstinence is limited to certain marginal social groups which are thus excluded from the political decision-making. A different story is when speaking of in the 'new democracies' where only recently the dissolution of authoritarian rule and the introduction of a democratic system took place, because in this case, the low turnout means distrust of a large part of citizens in consolidating democracy.

Voting is not the only form of political participation in modern democracy – although it is the most common and also the most important one. There are also other forms of citizen's participation in decision-making processes. There is engagement in political relevant organisations like political parties as the main institutional actors in modern political life as well as more unconventional groups, such as different social movements, ad-hoc associations (established to deal with certain specific social problems), 'protest' organizations, etc. Active aspect of such participation represents the practicing of certain politically motivated activities (such as political agitation) and engagement in various public manifestations (demonstrations, strikes, petitions). We can say that these forms of political participation are more complex since they require more personal commitment and engagement from the citizen.

These forms of political participation usually represented a useful complement to formal political processes. It is mainly about the development of the so-called civic competence which means the ability of citizens to play an active role vis-a-vis governing institutions in terms of participation in political decision-making and control over the conduct of power-holders (Soltan 1999, p. 17). Participation in various public organisations enables citizen's development of those attitudes and skills that are needed for successful governance. This refers primarily to the ability of argumentative political communication (both in terms of convincing their fellow citizens and in terms of defending their positions and proposals before the authorities) and various organisational skills needed to carry out certain political projects. Furthermore, public engagement of citizens contributes to strengthening the sense of community and thus the sensitivity of the issue of common good that increases social cohesion as an important component of integration of society.

These forms of participation are therefore related to public engagement in the sphere of politics. However, many critics of modern democracy address the fact that the participation of the citizens is confined to the political sphere. Some have also highlighted the need to supplement participation in political decision-making in the form of citizen participation in other areas in which they are directly engaged. In doing so, they most frequently mentioned the participation in the economic sphere, i.e. the engagement of employees in decision making within an enterprise in which they are employed. Such criticism came mostly from the ranks of Marxists and other radical theorists, which is understandable given the fact that Marxist social theory attributes the central role in society to the economic sphere. However, such criticism isn't confined only to these circles. For example, in his later works, Dahl (1989) claims that the concentration of power in certain areas – mainly economic – puts certain individuals and groups in a privileged position that negatively affects the quality of democratic decision-making. Therefore, it deals with the possibility of the application of democratic principles to decision-making processes in business corporations. The associativist theory advocates the extension of participation and democratic decision-making principles to the majority of areas of society, such as economic, social care, education, etc., since this is the only way to achieve the optimal functioning of these social areas (Hirst 1994).

It is true that participation in the workplace as well as other forms of people's participation in decision-making is important for the quality of life in the society. At least in the context of a social market economic model which is typical for continental Europe, employee participation is an established norm of economic/

industrial relations. However, the principles of political participation cannot be easily transferred to other areas of society – even (or even mostly) to the economic field. The functional goals of the various areas may be distinguished and thus the principles of their operation. The function of the economic field which is pursued by its institutional actors, i.e. the companies, is maximising of profits and their investment in the long-term development. This can be achieved only on the basis of rational organisation and optimization of resource which is not possible without expert knowledge, on the basis of which strategic decisions are made. Thus the criteria and mechanisms for recruitment to the key decision-making positions in politics and economics differ: in the first case, the relevant criterion is the support civic of the body, while in the second case, organisational competence is the most relevant.

In one stream of democratic political theory, participation of the citizens was not accepted as something univocally desirable. In fact, many scholars warned against the danger of extremist ideas and the spread of political radicalism, leaning on the experience of 20. and 30. years of the 20th century when a large part of Europe was overwhelmed by both right and left-wing anti-democratic movements. According to Huntington (1969, p. 79) excessive political participation in the case of weak political institutions can lead to political instability. Berelson, Lazarsfeld and McPhee (1955, p. 314) have even argued that a certain degree of political apathy is positive for political stability and for the successful functioning of democracy, as the more aggressively expressed political interests ought to be related to political extremism which is incompatible with democracy, while more moderately expressed interests allow political compromises. It is certainly true that not every type of political participation is desirable from a democratic point of view. This can certainly be said for engagement in activities of those organizations and groups that seek to enforce anti-democratic policy objectives, usually on the undemocratic way by undemocratic means. Therefore, it is necessary to distinguish between such illegitimate forms of participation and those who respect democratic principles - the latter in modern societies certainly prevail. These include those forms of political participation which do not take place through established political channels (i.e. through the institutions of the system) but their 'content and form' are not in conflict with democratic norms. (In this way, a protest in a public place - for example, before the parliament or the government - against a certain decision, made by the state authorities, is an unconventional form of participation but it is undoubtedly legitimate from a democratic point of view).

3. E-democracy and e-participation

E-democracy is related to the process digitalisation of decision-making processes in the context of democratic polity (Medimorec et al. 2011). This concept "refers the use of information and communication technology (ICT) in political debates and decision-making processes, complementing or contrasting traditional means of communication, such as face-to-face interaction or one-way mass media" (Päivärinta and Sæbø 2006, p. 818). Since democracy rests on the active engagement of the citizenry, its digital upgrading would involve the introduction of new participative tools, based on development in the realm of technology. E-participation can be perceived as the usage of ICT in order to enhance and deepen the political participation of citizens (Macintosh 2006). The main aim of e-participation is "to promote active citizen engagement in democratic societies and, through this engagement, to enable citizens to impact and foment change in governments and public policies" (Reddick and Norris 2013, p. 454). As such, it can encourage two-way communication between government and citizens, educate citizens about policy, legitimize government decisions, and provide opportunities for mutual learning (Phang and Kankanhalli 2008).

Certain authors stress that e-participation can have an impact a form of improvement of the public services, not including transformation of the nature of democratic political life (Reddick 2011). The participative model of interactions between citizens and decision-makers is very difficult to achieve since it involves changing existing power relationships (Kraemer and King 2006). Some of them also stress high costs of e-participation (Anderson et al. 2007). Introduction of ICT in the political sphere is more often related to e-Government as a tool for modernisation of public administration (Dobnikar and Žužek-Nemec 2007). This refers to availability of different online tools serving to needs of the citizens and companies like e-taxes, different e-registers and registration, electronic information databases, etc. These tools shall upgrade the quality of public services and thus increase satisfaction of its users.

Although improvement of public services is very important for the well-being of the people, it does not represent the essence of democratic life. What is more important is the activation of society through the introduction of participative tools that would compensate the decrease of traditional forms of political participation like voting turnout, membership in political parties and other organisations etc. ICT tools can be instrumental in this context in terms of enabling more direct communication between citizens and decision-makers and thus stimulating involvement of the first in political life on more individualised basis (e-initiatives, e-petitions, etc.). They can serve for their mutual informing on crucial issues that are a matter of public interest.

Different ICT applications can be also used to enhance political competence, perceived as the ability to understand political/democratic system, its institutions, agents and processes. Democracy needs competent and critical, well-educated and demanding citizens. There are tools of matching of political preferences (as we will see it the case of EUVox platform). On their basis, voters can examine their own policy preferences and better inform themselves about the orientations and preferences of political parties and their candidates when making their voting choice. This could, at least in the long run, lead to stronger bottom-up pressure on political decision-makers to make clear policy commitments that would enhance the possibility to hold them accountable after elections.

4. The state of e-participation in Central and Eastern Europe

According to the UN e-participation index for the year 2010, some of Central and Eastern European countries such as Estonia, Lithuania, Slovenia, Croatia and Hungary are listed among the first 30 countries in the world while the other countries of this region are ranked lower (Institute for Electronic Participation, Web). This implies that the level of e-participation in 'new democracies' varies from country to country. Estonia is undoubtedly the leading country in the Central and Eastern European region due to a considerable investment into high technology and its cuts across all sectors of the country (CEE Citizens Network: *E-Participation and e-Governance from the Perspective of Citizens in the Central and Eastern European Region*, Web).

An online consultation entitled *eParticipation in Central and Eastern Europe* in 2010 exposed the need of Central and Eastern European countries for sustainable eParticipation funding and a stronger civil society with their capacity of attracting resources (both financial and human) for e-democracy projects (Institute for Electronic Participation, Web). Furthermore, the strategic preparation for the development of eEurope until 2020 clearly shows the exceptional importance of adjusting e-participatory tools to the circumstances of the society and highlights as well the support of the authorities for reaching the effective e-participation (Pičman Štefančič 2008, p. 90). Regarding categorization of levels of e-participation[1], we

---

1 OECD (*Citizens as Partners: Information, Consultation and Public Participation*, Web) divides e-tools into three groups: information, consultation and active participatory. Pippa Norris (2004) divides e-tools into information, communicational and action tools.

focus on three types of e-participation existing in Central and Eastern Europe: information sharing, communicational and active participation in policy-making. Those represent three key democratic stages of e-participation which are indirectly interrelated in order to enhance the democratic participation.

Information sharing means both dissemination and consummation of information. It includes general information about the state and public institutions, policy documents and related matters, legal acts (both drafts and accepted legal acts), political public debates, campaigns and possibilities of civic participation (ibid. 57–62). Information is a crucial precondition for communicational and active participation. From the technological view, this type of democratic participation is one of the main mechanisms providing citizens easy and quick access to all the relevant information. There is a lack of interaction between government and citizens on information – one of the participants remains inactive. It is more or less a one-way approach. However, it represents a good starting point for the increase of democratic interest.

Communicational participation indicates active participation of both government and civil society, but the latter is not an equal participant in the decision-making processes. This category of e-tools requires the activation of participants but it does not predict a direct correlation between this operation and the final decision (ibid. 28). At this stage, the inclusive and active two-way interactions are provided. It is about the discussions and connections using the communicational tools that enable the citizens to share their opinion and positions about the political issues. The most common applications of communicational participation are e-forums, e-consultations, and e-petition.

E-forums that are supported either by the state or by the civic society[2] encourage the citizens to express and to exchange their opinions. It is a tool that allows citizens an exchange of individual and other views and opinions about a public matter. The aim of this tool is to strengthen the process of creating civic views through a deliberative confrontation which can take place before the political decision-making, after it or independently of it. On their websites, institutions of public authority can provide through this tool the opportunity for the exchange of opinions and views of interested citizens (Kukovič and Haček 2014, p. 32).

An example of this is the On-line Citizens forum European debates (http://www.evropske-razprave.si) as a Slovenian moderated online forum, providing and supporting the moderated online debates and online consultations between Slovenian members of European parliament and Slovenian citizens about the

2 There are different opinions about who should e-forums be organized by.

issues relevant for quality of life. The project is supported by the European office of European Parliament in Ljubljana.

According to the United Nations e-government survey in the year 2014, governments tend to use social media tools in a more extensive way than online forums. Social networks are growing fast and gaining many new users every day. The public is familiar with social media platforms such as Facebook and Twitter. Furthermore, social media initiatives do not require high investment costs in comparison with commercial and non-governmental platforms (United Nations: *E-government survey 2014*, Web, p. 70).

The authorities try with the use of different means of communication through e-consultation to cumulate the positions, views and preferences of the citizens and include them into the decision-making process. In opposition to e-forums, where both authority personnel and citizens may initiate different topics of interest and interact with each other, the e-consultation refers to the organized intention to get citizens feedback on specific issues (Panopoulou et al. 2013, p. 80). This tool represents a communication between citizens and authority as well as an interactive form of involving citizens into democratic process. There have been some e-consultation tools developed, for example e-poll and questionnaire as simpler but quite popular e-consultation tools.

E-petition does not belong neither to the communicational participation nor has the direct effect on the decision-making process (Pičman Štefančič 2008, p. 76). Regarding the fact that the pre-stage of e-petition is consultation and the e-petition does not have necessarily the mandatory effect, we define this tool as a form of communicational participation. In this regard, the citizens are eligible to initiate the petition on public issues and to support and sign the current petition. Citizens initiate the e-petition by adding content and other material to the online form. Therefore, other citizens can read the petition and electronically support it. Despite the strengths of the e-petition as simple usage, easy implementation and dissemination as well as low costs, formal e-petitions are not widely spread in new democracies.

An example of a communicational participation tool, we can state the Slovenian online tool *Predlagam vladi – I suggest to the government* (The Republic of Slovenia: *Predlagam.vladi.si*, Web) which is run by the Government Communication Office of Slovenia in order to provide an exchange of the positions, views and opinions about the political issues. The citizens set the proposals on relevant issues and therefore the assessment of these proposals could be conducted by the users (they can vote for them). The approved proposal could be sent to the body of government.

Active participation in policy-making represents the type of e-participation in which citizens interact with the decisions-makers and actively involve themselves in defining the process and content of policy-making. This type provides sufficient time and flexibility to allow the ideas and proposals on the part of citizens in order to integrate them into the policy-making process (OECD: *Citizens as Partners: Information, Consultation and Public Participation*, Web, p. 23). The responses are expected by the authorities who should in most cases mandatory follow the expressed will of the citizens. Active participation in policy-making includes regulated forms of e-elections and e-referendum.

Elections are placed at the heart of the democratic process therefore *e-elections* represent the fundamental e-democratic tool in order to have an efficient democracy system consistent with technological development[3]. By e-elections the citizens can choose those representatives of public authorities or officers who need to be elected. There are available also the additional mechanisms for online voter registration and other necessary operations (Kukovič and Haček 2014, p. 32). E-elections as time-saving and simple tools try to increase participation in decision-making process. It is reasonably that e-elections stimulate at first the participation of younger people due to their good technological education and flexibility.

The situation in the region in terms of availability of channels of e-participation improved in the last decade. Some countries actually became the front-runners in term of introduction of some of these channels. This particularly applies to Estonia which became the first nation that carried out general parliamentary elections over the internet in 2007 (preceded by the pilot project for the municipal elections in Tallinn in 2005). The electronic voting system was considered as a success by Estonian election officials. As mentioned above, Estonia is usually the highest-rated country in Central and Eastern European in different surveys on e-government; in general, countries from the region rank significantly lower than countries from Western Europe, North America and East Asia (*UN E-government Survey, Waseda University International e-Government Ranking*). We can claim that; speaking of e-democracy and e-participation, they lag behind developed Western democracies.

There are some projects in different countries of the region that use online technologies to promote accountability and transparency. *Technology for Transparency*

3 The regions of Central and Eastern Europe generally enjoy high internet levels penetration: in 2009, 70 percent of this area's population was online (Medimorec et al. 2011, p. 6).

Network states several initiatives that seek to increase enhanced openness and transparency of political processes as well as responsibility and accountability of their agents by informing citizens on their conduct, like *Sejmometr* and *Mam Prawo Wiedziec* in Poland, *Képmutatás* in Hungary, *KohoVolit* in the Czech republic and *Fair Play Alliance* in Slovakia.

However, e-participative tools do not only serve the citizens and civil society organisation to control power-holders but can be utilised also by political (and other) elites. In the basis of analysis of the use of such tools (like the voting assistance tool *Manobalsas* that was developed for the Lithuanian parliamentary elections of 2008 by three leading universities in Lithuania), some authors claim that parties might be able to manipulate the results of the tool by adopting non-ideological (ideologically unconstrained) populist policy positions (Ramonaite 2010).

In general, the informational component of e-democracy is much better developed than in the communicational one. A tool for informing citizens is introduced both the national and local level. The significant shift in responsiveness of the civil servants to the online requests occurred in the last decade (Pinterič 2007). There are different administrative online services available to them like tax reports and assistance (e-taxes), online car-registration, submission of project proposals, etc. On the other hand, there are very few tools for interactive communication of citizens with public officials and they are weakly developed, and even when existing, they do not perform the function of aggregation of articulation of common interests and are not a reflection of the issues of common relevance (Lukšič and Delakorda 2007). There are on-going debates on the possibility of introduction of the online voting, however, the political climate is still not very favourable for this type of participation.

Despite the fact that the Central in Eastern European countries made a great move forward in the terms of strengthening e-participation during the last years, in comparison with West European countries, many e-participation tools remain in the preliminary stage of development. The CEE Citizens Network Survey (2013) indicates a considerable interest among respondents for e-participation and collaboration but it emphasizes as well some problematic points. This is related to the general level of economic and technological development but also to the nature of institutional setting. It is a rather hierarchical governance structure, coupled with a rigid administrative culture in these states that can hardly facilitate stronger contributions by their citizens.

5. E-democracy tools for enhancing political participation and competence – the case of EUVox

The increasing use of the internet has influenced society and indirectly required the inclusion of much information in the internet database. One should take into consideration the fact that the internet has become a new source for obtaining information about political parties and candidates. Therefore we could enhance participation in elections and the running of the democratic process by better politically informed voters. There were several e-democracy tools that were designed to help voters to detect their political profile vis a vis those who compete for their votes, i.e. political candidates/parties. One such example was the EU Profiler, developed for elections for the European Parliament in 2009. The most recent tool of that kind is the EUVox which provides voters the essential information about the positions of the political parties and stimulates political debate. More broadly we are talking about the platform so-called Voting Advice Application – (VAA)[4].

EUVox 2014 is an independent online application sponsored financially by the Directorate-General for Communication of the European Parliament and Open Society Initiative for Europe. The project was designed to compare the positions of EU citizens and major political parties of each EU country participating in the election to the European Parliament. It was developed by a consortium of researchers based on *Kieskompas* and *Preference Matcher*[5] in collaboration with researchers with expertise from all 28 member states of the EU. The application based on different methodologies helps the users to determine how their preferences match the preferences of political parties (EUVox (1) 2014, Web).

The EUVox application is divided into two main parts. The first part is designed as a classic online questionnaire about political views and positions that all the voters can fill in. It is formulated in approximately 30 questions gathering three essential political contents: European integration, economy and social policy (same-sex marriage, abortion, immigrations…). The second part of the EUVox is identified by the questionnaire that coordinates the positions of the main political parties of all EU countries on the political issues. An independent group of European political experts defines the beliefs and positions of the parties based on the parties' programmes and official documents, the statements of the parties' representatives, the moves of the parties up to this point, opinions of outside

---

4 VAA's are online databases of parties' positions on a number of core policy issues (Cedroni and Garzia 2010, p. 9).

5 Kieskompas and Preference Matcher are experienced organizations in the field of VAAs that have designed many such tools across four continents (EUVox(1) 2014, Web).

experts. (EUVox(2) 2014, Web) According to the answers of the voters, the result of matching is calculated and graphically[6] represented by the application. Hence it is possible to estimate the closeness of the voter to the various political parties.

The EUVox application does not give the answer to the voter which party they should vote for but it shows them how their positions fit to the main parties across a range of policy issues (EUVox(2) 2014, Web). The system enables the voters to get information about the views of political parties and encourage them to reflect on the political issues relevant for the elections to the European parliament. The application serves as an aid to the people to orient themselves in the political space as well as a tool for highlighting the current political issues in the electoral campaign in order to enhance the confidence of the voters and stimulate electoral turnout.

Data on participation in the EUVox survey show big differences across the European Union. There are some countries with traditionally high numbers of those who take part in such activities (this holds especially for Scandinavian countries). Although participation in CEE countries was, on average, lower than of the 'old' EU members, in some of them people expressed rather high interest in it (for example, the Czech Republic had almost twice as many EUVox participants as Austria which is comparable with it in terms of the population).

According to the data of the European Commission, the participation in the elections to the European parliament has been declining generally probably due to remote, complex matters and key issues of the European Union towards attitude of the voters (European Commission(2): *European Parliament elections – getting out the vote,* Web). Eurobarometer (European Commission(1): *Eurobarometer survey #364 on electoral rights,* Web) found that 84% of citizens in 27 EU member states agree that voter turnout would be higher if more information was provided on the programmes and objectives of candidates and parties in the European parliament. The EUVox was designed to diminish the effect of lack of information about parties' affiliation. Clarifying the entire picture of the political space in each particular country, the EUVox could serve to raise the participation in the elections to the European parliament.

It might be doubtful whether the competition of political parties in 'new democracies' is based on their programmes and political directions. There is a trend of 'personalisation of politics' as expressed in increasing impact of personalities

6 The diagram shows the matching between parties' and voter's positions and it ranges from – 100 (total disagreement) to + 100 (total agreement). Scores between 0 and 40 represent a weak match between a party and a voter (yellow bar). Scores about 40 or above mean a strong match (green bar) and scores less than 0 shows a negative match (red bar) (EUVox(2) 2014, Web).

of the parties' representatives on the electoral outcome (Tomšič and Prijon 2012). We cannot neglect the influence of mass media as one of the substantial factors of electoral campaigns. People often do not judge political parties and their candidates predominantly by their programmes or ideas. The choice that people will vote for is due to so-called 'cult of personality'. Online application EUVox could thus boost the impact of party's programmes on decisions of voters, prevent the personalisation of voting behaviour and stimulate the participation in the elections to the European parliament.

6. Conclusion

The global economic crisis corresponds with the decreasing of trust in political institutions and the political elite as a whole (Tomšič and Prijon 2013). Many protesters that appear all around Europe who express their revolt against political elites and their misconduct demand, among others, more directed participation of citizens in the political process, believing that it would enhance accountability of decision-makers. Enhanced implementation of mechanism for e-participation can play positive role in this regard.

Political life in CEE countries is still characterised by relative instability and volatility. This is mostly a consequence of weakly profiled identity of many political parties. Furthermore, it is characterised by a number of elements that are related to the misbehaviour of political elites in the form of incompetence, clientism, corruption, ideologisation. This leads to widespread apathy among citizens, i.e. their lack of willingness for political and civic engagement as expressed by low voter turnout and party membership. In such circumstances, political populism can rise. Political leaders with excellent rhetorical skills, effective communication styles and personal charisma can spread their demagogic appeals, promising easy solutions even for the most complex social problems. Populism of either leftist or rightist ideological orientation is detrimental from the perspective of development of democracy since it can easily lead to political and social exclusivism. For populist approaches, it is very unlikely that they would be able to tackle challenges these societies are facing in a productive way.

In order to avoid populism and other threats to democracy, active and well-informed citizens are necessary. The availability of ICT tools for e-participation is very beneficial in this regard. Furthermore, there is an urgent need for strengthening citizens' participation and competences by promoting the public knowledge about e-democracy and its transparency. In order to increase e-participation, we have to also stress the importance of the responsiveness of authorities to the citizens who actively use the e-participative tools. Without it, political apathy would not wither away.

7. References

Andersen, Kim V. / Henriksen, Helle Z. / Secher, Christine / Medaglia, Rony: „Costs of e-participation: the management challenges". *Transforming Government: People, Process and Policy* 1, 2007, p. 29–43.

Beck, Ulrich / Beck-Gernsheim, Elisabeth: *Individualization*. Thousand Oaks: London, SAGE Publications: New Delhi 2002.

Berelson, Bernard / Lazarsfeld, Paul / McPhee, William: *Voting: A Study of Opinion Formation in a Presidential Campaign*. University of Chicago Press: Chicago 1955.

Berlin, Isaiah: *Four Essays on Liberty*. Oxford University Press: Oxford 1969.

Cedroni, Lorella / Garzia, Diego (eds.): *Voting Advice Applications in Europe: The State of the Art* 2010, retrieved 25.02.2015, from https://www.academia.edu/281395/Voting_Advice_Applications_in_Europe_The_State_of_the_Art.

CEE Citizens Network: *E-Participation and e-Governance from the Perspective of Citizens in the Central and Eastern European Region*, retrieved 16.02.2015, from http://unpan1.un.org/intradoc/groups/public/documents/un/unpan023683.pdf.

CEE Citizens Network: *Results from CEE CN survey on electronic participation. Do you e-participate?* Prepared by Simon Delakorda, Ljubljana 2013, retrieved 05.02.2015, fromhttp://www.ceecn.net/images/stories/2013/cee_cn_e-partici pation_survey_report_final.pdf.

Centre of eDemocracy: *E-participation*, retrieved 13.02.2015, from http://www.inepa.si/e-participacija.si/en/center-za-e-demokracijo.html.

Dahl, Robert A.: *Polyarchy: Participation and Opposition*. Yale University Press: New Haven, London 1971.

Dahl, Robert A.: *Democracy and its Critics*. Yale University Press: New Haven, London 1989.

Dahl, Robert A.: *On Democracy*. Yale University Press: New Haven, London 1998.

Dalton, Russell: "Value Change and Democracy". In: Pharr, Susan / Putnam, Robert (eds.): *Disaffected Democracies*. Princeton University Press: Princeton 2002.

Dalton, Russell: *Democratic Challenges, Democratic Choices*. Oxford University Press: Oxford, New York 2004.

Dobnikar, Aleš / Žužek-Nemec, Nataša: "E-Government in Slovenia". *Informatica* 31, 2007, p. 357–365.

European Commission(1): *Eurobarometer survey #364 on electoral rights*, report on 03/3013, retrieved 16.02.2015, from http://ec.europa.eu/public_opinion/archives/flash_arch_374_361_en.htm#364.

European Commission(2): *European Parliament elections – getting out the vote*, 13.3.2013, retrieved 16.02.2015, from http://ec.europa.eu/news/eu_explained/130313_en.htm.

EUVox(1) 2014, retrieved 15.02.2015, from http://www.euvox2014.eu/.

EUVox(2) 2014, retrieved 15.02.2015, from http://www.euvox.eu/si/questionnaire/faqs.php?lang=english.

Hirst, Paul: *Associative Democracy*. Polity Press: Cambridge 1994.

Huntington, Samuel: *Political order in changing societies*. Yale University Press: New Haven, London 1969.

Inglehart, Ronald: *Modernization and Postmodernization: Cultural, Economic and Political Change in 43 Societies*. Princeton University Press: Princeton 1997.

Institute for Electronic Participation (INePA), retrieved 15.02.2015, from http://www.inepa.si/index.php.

Kraemer, Kenneth / King, John L.: "Information technology and administrative reform: will e-government be different?". *International Journal of Electronic Government Research* 2 (1), 2006, p. 1–20.

Kukovič, Simona / Haček, Miro: "The Distribution of E-democracy and E-participation Tools in Slovenian Municipalities". *World Political Science Review* 10 (1), 2014, p. 25–44.

Lukšič, Andrej / Delakorda, Simon: *Kako demokratizirati e-demokracijo na državnem portelu e-uprave* 2006, retrieved 15.02.2015, from http://www.e-participacija.si/files/kako_demokratizirati_demokracijo.pdf.

Macintosh, Ann: "eParticipation in Policy making: the research and the challenges." In: *Exploiting the Knowledge Economy: Issues, Applications and Case Studies*. IOS press 2006, p. 364–369.

Makarovič, Matej / Tomšič, Matevž: "Democrats, authoritarians and nostalgics: Slovenian attitudes toward democracy". In: Jurak, Pandiloska / Pinterič, Alenka and Uroš (eds.): *Contemporary world between freedom and security*. Vega: Ljubljana 2010, p. 461–487.

Medimorec, Daniel / Parycek, Peter / Schossböck, Judith: *Vitalizing Democracy through E-Participation and Open Government: An Austrian and Eastern Perspective. BertelsmannStiftung* 2011, retrieved 08.02.2015, from http://www.fundacionbertelsmann.org/cps/rde/xbcr/SID-853E53E9-9B1F1F76/bst/Daniel%20Medimorec.pdf .

Norris, Pippa: *Deepening Democracy via e-Governance. Draft chapter for the UN World Public Sector. Harvard University* 2004, retrieved 14.02.2015, from http://www.hks.harvard.edu/fs/pnorris/Acrobat/World%20Public%20Sector%20Report.pdf.

OECD: *Citizens as Partners: Information, Consultation and Public Participation* 2001, retrieved 14.02.2015, from http://www.ecnl.org/dindocuments/214_OECD_Engaging%20Citizens%20in%20Policy-Making.pdf.

Päivärinta, Tero / Sæbø, Øystein: "Models of E-Democracy". *Communications of the Association for Information Systems* 17 (37), 2006.

Panopoulou, Eleni / Tambouris, Efthimios / Sanchez-Nielsen, Elena / Zotou, Maria: "Learning from eParticipation initiatives of regional and local level authorities in Greece and Spain". In: Loukis, Euripidis / Macintosh, Ann / Charalabidis, Yannis (eds.): *E-Participation in Southern Europe and the Balkans: Issues of Democracy and Participation Via Electronic Media.* Routledge: London 2013.

Phang, Chee W. / Kankanhalli, Atreyi (2008): "Framework of ICT exploitation for e-participation Initiatives". *Communications of the ACM* 51 (12), 2008, p. 128–132.

Pičman Štefančič, Polona: *E-demokracija.* Uradni list Republike Slovenije: Ljubljana 2008.

Pinterič, Uroš: "E-Governance in Slovenia". *Public Manager* 35 (4), 2007, p. 61–64.

Putnam, Robert D. / Pharr, Susan J. / Dalton, Russell J.: "Introduction: What's Troubling Trilateral Democracies". In: Pharr, Susan J. / Putnam, Robert (eds.): *Disaffected Democracies.* Princeton University Press: Princeton 2000, p. 3–30.

Ramonaitė, Ainė: "Voting Advice Applications in Lithuania: Promoting Programmatic Competition or Breeding Populism?" *Policy and Internet* 2 (1), 2010, p. 117–141.

Reddick, Christopher: "Citizen interaction and e-government: evidence for the managerial consultative, and participatory models". *Transforming Government: People, Process and Policy* 5 (2), 2011, p. 167–184.

Reddick, Christopher and Donald F. Norris: E-participation in local governments. *Transforming Government: People, Process and Policy* 7 (4), 2013, p. 453–476.

Smith, David Horton et al.: *Participation in Social and Political Activities.* Jossey-Bass Publishers: San Francisco 1980.

Soltan, Karol E.: "Civic Competence, Attractiveness and Maturity". In: Elkin, Stephen L. / Soltan, Karol. E. (eds.): *Citizen Competence and Democratic Institutions.* The Pennsylvania State University Press: University Park 1999, p. 17–37.

Tarrow, Sydney: "Mad Cows and Social Activists: Contentious Politics in the Trilateral Democracies". In: Susan Pharr J. / Putnam, Robert D. (eds.): *Disaffected Democracies.* Princeton University Press: Princeton 2002, p. 270–310.

The Republic of Slovenia: *Predlagam.vladi.si,* retrieved 09.02.2015, from http://predlagam.vladi.si/.

Tomšič, Matevž / Prijon, Lea: "Personalized politics and political communication: the case of Slovenia". In: Brusati, Luca / Fedele, Paolo / Iannello, Mario (eds.): *Economia e comunità: cinque anni di ricerca al servizio del territorio.* Forum: Udine 2012, p. 129–145.

Tomšič, Matevž / Prijon, Lea: "Elites, ideologies and crisis in Slovenia". *Raziskave in razprave* 6 (2), 2013, p. 71–116.

Tomšič, Matevž / Vehovar, Urban: "Quality of governance in "Old" and "New" EU member states in a comparative perspective". *Sociológia* 44 (3), 2012, p. 367–384.

United Nations: *E-government survey 2014*, retrieved 15.02.2015, from https://publicadministration.un.org/egovkb/Portals/egovkb/Documents/un/2014-Survey/E-Gov_Complete_Survey-2014.pdf.

Victor Cepoi
School of Advanced Social Studies in Nova Gorica
Gregorčičeva 19, 5000 Nova Gorica, Slovenia
victor.cepoi@fuds.si

Trust and Participation in the Information Society: New and Traditional Information Sources

Abstract: *In this paper, the primary focus is on trust and sources of information in enhancing or hindering participation in Europe. More specifically, the Structural Equation Modeling is used to trigger the explanatory conditions for participation. Results from the latest data from the World Values Survey, show that in the case of participation, new and traditional forms of information sources are moderate explanatory factors. In the case of trust, only generalized trust can be highlighted, meanwhile political trust did not put forth any significant outcomes. In addition, the analysis showed the positive relation between trust and traditional information sources. Nevertheless, the proposed model underlined the contradictory effect of level of education as in comparison to what has been known until now.*

Keywords: *trust, participation, new and traditional information sources*

1. Introduction

With the development of the internet and the alignment of the modern societies toward informational ones, the old methods toward many political and sociological problems are facing a rethinking in terms of concepts and approaches. The new challenges have to be addressed considering the technological tools that influence the daily functioning of the society. The enormous incensement of Internet users in the last couple of years cannot remain without consequences for political participation and social capital. The Internet can represent a mean for the construction of the social capital (Gainous, Marlowe, and Wagner 2013). This argument relies on the increasing evidence of the effects of the online framework on the creation of social capital. For the reason that citizens in present-day societies are unwilling to participate in politics (Fuchs 2007) and the decline of political and generalized trust (Hardin 2006) has become a major worry for many observers. Trust and community stimulate political participation (Gainous, Marlowe, and Wagner 2013, p. 147).

In most of the democratic societies, trust has a big importance not only for people but also for the entire system (Mishler and Rose 2005). The phenomena of the presence or absence of trust is a much debated issue among social scientists. Because of the interaction between people on different levels, trust needs to be explained in the best possible ways. Low social capital manifests itself in many areas, such as low interpersonal trust, weakness of community spirit at a local level, weak trust in political institutions, and business practices that affect the social image of the private sector (Bartkowski 2003, p. 184).

In a society in which the level of trust is constantly declining, it is expected to have a decline also in other spheres of the daily life of the society, which are strongly linked to the concept of trust. Relying on this perspective it is very likely that there will be a domino effect on other components that form the social capital on a society. Furthermore, it will have an impact of the society and democratic consolidation of the regime and the way people are perceiving democracy and its consolidation. Social capital can be viewed as a combination of different elements, such as social relations within families, communities, friendship networks, voluntary associations, shared values, norms and habits, generalized and instructional trust, and civic morality (Oorschot, Arts, and Gellissen 2006). All of these dimensions will have an influence on the level of social capital in a society. As Oorschot et al. mentioned that social capital is more an aggregation of different variables, rather than an individual one. It is also seen as a terminological umbrella for grouping together a wide range of social phenomena. This means that we have several concepts that once aggregated will form the broader concept of social capital. The issue of trust is still an important one even in consolidated democracies, where trust in the last period has faced a decline in the society (Cohen 1999). The decline of trust, especially when it comes to institutional trust, is a common characteristic for both established and young democracies. It has to be noted that when it comes to institutional trust, it is perceived as the level of trust people have in state institutions (such as parliament, government, etc.), meanwhile the concept of political trust emphasizes a broader set of actors.

In this paper trust and participation in the aspect of technological development will be at the forefront. The main argument on which this paper is constructed is on the positive relationship between the three main concepts: trust, political participation, and internet use. For this concern, the theory of social capital and political participation, but also the role of internet will be the background of the further analysis in this paper. The latest data on an aggregate level from the World Values Survey was chosen for detailed study to explain political participation in 12 countries from Europe (both EU and non-EU members). The theoretical considerations will be tested through the Structural Equation Modeling, which shows potential causal dependencies. Moreover, the focus on the casual relationship can be specified either on hypothesized relations or already estimated ones (based on previous researches).

2. Forms and the Importance of Trust

Trust is the core of social capital, which is an important factor for modern societies (Freitag and Bühlmann 2009). Moreover, it is argued that trust can affect the way a regime is perceived by the society. Not only that the regime could lose its legitimacy, but also people become less trustworthy, which has an impact on social cooperation, and makes the society lose its cooperativeness. Scholars tend to assume that trust might be the cause/ effect or both of civic engagement, because both concepts have tangled (Uslaner 2007, p. 152). The rich empirical research that has been carried out by scholars regarding political participation, has contributed to the enrichment in findings concerning the variations, types or amount of public participation in political life (Finkel 1985).

Confidence can be considered as the foundation of solidarity and governments that are more effective, higher growth rates, less corruption and crime. The author argues that the level of trust depends on information and experience (Uslaner 2007, p. 153). On the other hand, distrust can play a vital role for the mobilization process. Distrust of the systematic level can lead to the challenge, debate, and struggle stagnant and exclusionary political formations (Mische 2012). It could generate institutional innovation, contestation, and engagement, things that describe a strong democracy. Schyns and Koop (2009), in their analysis of the effect of political distrust on social capital, found that the effect of political distrust on citizenship norms is not very strong. Both trust and distrust are important for democracy, even if at different scales.

Dahl (1999) pointed, as a paradox to the presence of distrust, that democracy is still "the only game in town" in many democracies. For large-scale democratic government within societies we speak about two dimensions of democracy. When we refer to the first dimension, political institutions are important. The role of these actors is to create the rights and opportunities for the citizens within the society. The first dimension is the level of trust that citizens have in the political institutions that regulate the state activity. In the moment citizens will have a high level of trust, they will start to be more civic engaged. The second dimension is referring to the participation of citizens (Dahl 1999, p. 13). It is not enough to have rights and opportunities, but rather that these means must be used in order to have a strong democratic regime. Political institutions and everything that is related to the first dimension of democracy (even the trust in these political actors) can be regarded as a tool for the well-functioning of the society.

Trust is a vital element in different levels of cooperation such as social, economic or even political, with different type of actors, for instance family, friends or even strangers. In modern societies, trust is one of the key elements of a good and productive coexistence of the entire society (Freitag and Bühlmann 2009, p. 1538). Perhaps trust is the most important component of social capital, which is a necessary condition of social integration, economic efficiency, and democratic stability (Newton 2001). As Uslaner (1999) stresses, the presence of trust or other forms of social capital within the society depends on the society's environment. More specific, the presence or absence of the democratic order influences trust. Within democratic regimes, people are more trustworthy, because democracies promote criticism and variety. On the other hand, a more trusting political culture is more conducive to democracy. The primary argument is that the more trusting a society is, the more tolerant and more accepting of minority cultures. Nevertheless, the small amount of empirical studies regarding the relationship between the two, determines us to focus on the symbiotic relationship, where democracy is important for trust and trust is important for democracy.

Freitag and Bühlmann (2009, p. 1538) mention that trust can be perceived on two different levels. Scholars distinguish between particularized trust and general trust, which means having confidence in those people that you have not met and do not know about their trustworthiness. Particularized trust is being characterized by having trust in other people that are making part of their own group. Generalized trust is being characterized as willing to trust strangers that not necessarily are like themselves, and consider that by making contacts with different people will result into beneficial positive results.

Trust in institutions can be understood as the confidence of people have in the political environment (Catterberg and Moreno 2005). The problem of a low level of political trust is not only in new established democracies, but also in consolidated democracies, which are suffering a long-term crisis, and worry that the democratic regime could eventually collapse (Blind 2006). There are two broad approaches when it comes to the explanation of political trust. Firstly, we have the institutional approach, which is based on the rational choice perspective. Secondly, there are the cultural approaches, which in contrast to the institutional approach scholars view it as originating not in the political sphere, but outside of it, in the cultural values and beliefs of the people (Mishler and Rose 2001).

Most of the studies in Western democracies focus on government institutions and emphasize the performance of incumbent leaders rather that look just at the institution, which according to them it is an important exception (Mishler and Rose 1997, p. 425). Their study has shown that citizens in the post-Communist countries of Eastern and central Europe have a median attitude of the institutions of the civil society. One of the main problems with this concept is that the literature has not distinguished between confidence in the institutions on the representational side of political system, represented by parties, parliament, etc., and on the implementation side of the political system (Rothstein and Stolle 2008). There is a theoretical reason to make this difference, because on the representational side, the main role of the political institution is to be a partisan (Rothstein and Stolle 2008, p. 9). Meanwhile, on the implementation side of political system a party that holds government power or a majority in the Parliament will want to implement its ideology.

The main difference between social and political trust is that social trust is based on the personal knowledge of other persons meanwhile political does not imply this type of knowledge (Newton 1999). These two types of trust belong to different spheres, though this thing does not necessarily mean that these spheres are separated. Personal trust belongs to the private sphere, but the political trust to the public sphere. Both types of trust are more dependent on formal sanctions in the shape of rules and regulations about the conduct of politicians (Newton 1999, p. 171). The author mentions, that political trust is strongly associated with political differences. It means that there are those who identify themselves with the government. In addition, there are people who identify themselves with the opposition. It depends with which political actors, with whom people identify themselves and what political actors are representing political institutions in the country. It is important to look also at this aspect, because it can explain the evolution of political trust that is present in the state.

The difference between these two types of trust consists in the agents toward whom trust is being oriented (Newton 1999, p. 170). In the case of generalized trust it refers to citizens that have confidence in other citizens, meanwhile political trust is referring to the level of trust that the citizens have in political institutions or in political leaders. Nevertheless, both of these types of trust are important for a democratic society, and for the democratic regime itself and for the concept of social capital.

3. Political Participation

One of the definitions for political participation is that it is characterized by taking part in public controversies, power struggles, and disputes. It regards the acknowledgement of differences and diversity, which makes political participation a contentious issue (Hedtke and Zimenkova 2013). Democracy, characterized by communitarian forms of involvement emphasizes the importance of political participation (Fennema and Tillie 1999). When it comes to democratic decision-making in the rational choice model and the tool for overcoming the collective action dilemma, political participation can overcome it by creating trust (Fennema and Tillie 1999, p. 703). Both the democratic systems and participation have an inbuilt tendency for disappointing its citizens, thus disappointment can be considered the main reason for non-participation (Hedtke and Zimenkova 2013, p. 3). This is possible when these systems stop working, therefore making society unwilling to participate, because of the lack of belief in changes. Citizens that actively participate in the public life develop an attachment to it, which promotes solidarity and participation of people in public affairs.

Political engagement can be understood as "activity that has the intent or effect of influencing government action" (Verba, Schlozman, and Brady 1995). The intent or effect can be either directly (affecting the making or implementation of public policy) or indirectly (by influencing the selection of policy-makers). Political participation includes activities such as working for a candidate or party, trying to convince someone how to vote, or working to affect the making or implementation of public policies by officials. However, voting is the most important activity (Keeter, Andolina, Jenkins, and Carpini 2006). Participation can occur on two levels: direct or indirect. Direct citizen participation is characterized by public debates, popular meetings, protests, referendums, etc. Meanwhile elections, parties or elected positions are considered as indirect forms of citizen participation (Aars 2007)

Generally, the explanation of political participation focuses on levels of education and income. For the reason that these models are limited, it has led to renewed interest toward this particular topic (McClurg 2003). If in case of the focus on the explanation of political participation on education or income research is big, when it comes to social interaction the investigation is relatively little.

4. A Digital Approach Toward Participation

The development of the Internet as part of the democratic process has stimulated the discussion regarding its role in civil and political processes (Tiscornia and Fernandez-Barrera 2012). It enables citizens to promote a more active role than before, by reducing the costs of the participation. The technological development in the sphere of information and communication considerably reduces the obstacles to the realization of participatory democracies (Fuchs 2007, p. 43). These technological possibilities formulate the basis of participatory democracy relying on electronic democracy. The virtual space can be regarded as an area for communication between citizens and the common will formed, which does not mean that the deliberation will not raise further problems (Fuchs 2007, p. 49). New forms of citizens engagement have both positive and negative effects on democracy (Karlsson 2012). On one hand as the author argues, citizens can more easily access the knowledge in the policymaking process and increase political trust. On the other hand, these forms of engagement can increase the political inequality. Some authors (De Marco, Antino and Robles Morales 2012, p. 62) do not consider digital political participation as part of offline participation, but as a form of participation in itself, mainly because of the absence of a convincing relation of influence between the two forms of participation.

With the development of the Internet and becoming more integrated with daily life, the role of social networks must not be omitted. Because of the decline of trust in institutions, public authorities and media, the fear is that people would turn to forms of online communication (Quandt 2012).

Trust and societal communication are related. Social interaction in the online sphere impacts the strategies and tactics of everyday life (Dahlgren 2012). It allows the horizontal communication between people and organization with the purpose of sharing information, organizing, mobilizing or providing support. Nevertheless, the important role of social media are not decisive on their own.

Another aspect of the overlap of technological development and democracy is the domain of e-participation. In order to improve the efficiency and legitimacy of the political process, governments seek to encourage the participation through different means. In present, societies have the possibility to engage and support the decision-making processes through different forms of e-participation, such as discussion forums, blogs or electronic voting system (Glassey and Aglione 2012). As a result, it can be argued that e-participation firstly reduces the "costs" of participation. More specifically, same traditional forms of participation can be embraced.

All the procedures are more accessible for people and allow people to embrace the roots of democracy. Nevertheless, results show that traditional channels for participation are more used than the e-participation ones, but also the reasons to participate remain rather traditional.

5. Trust, Participation and Internet

In the literature, the usage of the Internet can be considered as part of the digital skills (De Marco, Antino and Robles Morales 2012, p. 69). The level of Internet usage allows one to measure the the appropriation of digital skills of the users.

This line of reasoning can also link to trust and media and political information exposure. Moreover, even reading news on the internet can be regarded as stimuli for participation (De Marco et al. 2012, p. 71). People, who trust others, are in general the strongest supporters of the fundamental norms that make for a civil society and cooperative society (Uslaner 2002). Trust and one's own moral code lead people to approve standards of moral behavior. Tyler (2006) stresses that in the moment organizations are perceived as being accountable and act in an impartial manner, the level of legitimacy will increase among society. Additionally, it will increase their willingness to comply with the rules that are present in society.

It was confirmed that citizens who live in more trusting and civic communities, are socialized into trust, reciprocity and honesty. It means that not only they are more likely to develop the heuristics of trust than those who live in non-civic regions, but also they are more likely to respect the norms, rules and obligations of the community life (Letki 2006, p. 4). Firstly, there must be a trusting community for the case of the generalized trust. In the case of political trust there must be trusting political institutions. These two variables ought to result in the development of civic morality among citizens. In addition, it will affect the level of civic morality in the society, and we could see the future development of the social capital within the society. A person, who is trustful toward others, will be likely to develop positive attitudes toward a larger community (Listhaug and Rigndal 2004, p. 344). I hypothesize that in order for people to engage in political actions firstly they have to exert high level of trust. The bigger the level of generalized trust of a person, the bigger are the chances for this person to interact with others, and engage in different types of political actions. Moreover, the more people are dissatisfied with the current political elite, the more they are likely to engage in actions that can be considered as political participation.

Letki and Evans (2005) looked at the relationship between democratization (change in level of democracy over time) and social trust, and the orientation of this relationship. The authors were interested not only in the causality of the relationship, but also the direction of this relationship. Nevertheless, they found that the impact of democratization on social trust is very big. Freitag (2006) has found the empirical evidence in the degree of social trust has an impact on the social capital index. These researches show that there is a link between trust and social capital, and this link is very strong. This link makes the connection between concepts used to measure social capital and social trust also solid enough to have an impact on their development of change. On the other hand, it is considered that any type of political participation usually is very important for institutions and their credibility (Letki 2003). Moreover, citizens by not participating within these processes, increase the decline toward a civic community. Additionally, it is considered that in the moment a society is not trustworthy, it might cause a negative effect on participation within the democratic processes.

Nevertheless, in the modern informational society we must expect that the new available means of interaction could also affect the way people are interacting and the appearance of new explanatory models for political participation. Therefore, as the author concludes, social trust has a positive effect on voting, highlighting that low level of trust leads to low level of political engagement (Letki 2003, p. 24).

When there is present mutual trust between people, there will be reciprocity (Uslaner 1999, p. 6). When people take this approach as a moral value, they express a generalized faith in other people. As a moral value, it reflects the disposition to believe that others share the same values and norms as the society does. Furthermore, they will not act against others' interest even when they have the opportunity to do so, even in the case that they would not share the same specific interests at that specific moment of time. Social ties create the sense of community that reinforces values promoting cooperation. This means that when someone has trust in other citizens of the society, there is present a bigger cooperation between people (Uslaner 1999, p. 9). When one lives in a society with a high level of generalized trust, people tend to interact more and be more co-operative.

Generally, it can be stated that there is a difference between western and eastern societies when it comes to the analysis of civic engagement and trust (Uslaner and Conley 2003). One of the arguments sustaining this idea is that civic participation was not an option. Moreover, the differences in regimes work through the underlying factors for trusting others and civic engagement. Trust is linked to the involvement in the associational life. Following this line of argument, generalized trust can be considered as one of the resources for democratization processes and the stability of the democratic regimes (Badescu and Uslaner 2003). Additionally, in societies that are more trustful, citizens tend to be better at overcoming collective action dilemmas (Yamagishi and Yamagishi 1994).

As mentioned earlier internet usage not only diversifies the information sources, but also reduces the costs of the gathering process. Therefore it can be argued that the more accessible the information the more people willing will be willing people to inform or even participate in the political process (Verba, Scholzman, and Brady 2010). Moreover, it affects the way individuals perceive others during the deliberation process. In many cases people are just not aware of the other's status, which might benefit to produce more open and democratic communication (Bhuiyan 2004). Additionally, as the author stresses, using email instead of face-to-face meetings generated more proposals for discussions, but also contributed to the omission of any kind of differences. Stanley and Weare (2004), stress that the new technological developments impact participation by triggering the motivation (by having a qualitative effect on psychological perceptions of political information), capacity (reducing the costs of communication or collective action), and opportunities (opening the closed elite networks) of the society.

Nonetheless, there is no consensus regarding a sole answer about the presence or absence of a positive or negative relationship between the Internet (as a tool of information and participation) and actual participation. Therefore, I hypothesize that those who tend to use the Internet will tend to be more political involved.

6. Data and Result Interpretation

In order to test the proposed model, in the analysis data from the World Value Survey[1], was used because of several reasons, in particular avoiding differences in the methodology of the data collection. Nevertheless, when it comes to comparative studies, a limitation is that cultural, linguistic and other differences that might appear between them must be considered. Moreover, questions or acts regarding sensitive topics can produce social desirable responses. People may misrepresent themselves, because there is the possibility that they will tend to fit their responses within the framework of social norms, behaviors, or attitudes (Kreuter, Presser, and Tourangeau 2008). In addition, topics that may be disposed to the effects of social desirability effects range from abortion to religious service attendance, though some topics are more predisposed to social desirability biases (Kreuter, Presser, and Tourneau 2008, p. 848).

Discussing about **generalized trust** especially in the case of a comparative approach toward it, the validity of generalized trust measurements must be taken into consideration (Reeskens and Hooghe 2007). The problem relies within the question that is being used to measure generalized trust and in the instruments that are used to measure it. The indicators for measuring the latent concept generalized trust can differ from country to country though the measurement of the latent variable in a cross-cultural context can be done through different methods. Referring to the concept of generalized trust, the question used to assess the level of trust in a society is: "*Generally speaking, would you say that most people can be trusted?*" (Knack and Keefer 1997, p. 1256). What the authors mention is that there is ambiguity with what people mean, because it could measure the level of trust

1 The analysis focused on the available data from Europe, including the following countries: Belarus, Cyprus, Estonia, Netherlands, Romania, Russia, Slovenia, Spain, Sweden, and Ukraine

in close friends, relatives, or strangers. In addition, they can refer to people they most transact with, when dealing with this perception. For the proposed model, generalized trust will be constructed from two questions that measure the level of trust in people you meet for the first time and the agreement with the statement that most people can be trusted.

Looking at the support for regime institutions, we can include attitudes towards governments, parliaments, the legal system and police, political parties, and state bureaucracy (Norris 1999). It offers the possibility to measure and observe political trust not only in one particular actor, but how citizens' political trust developed during an amount of time. Furthermore, it offers the possibility to compare among actors, but also to make some assumptions on how political trust in evolving in country at a certain point of time. Referring to the political trust that a society offers to different actors, there can be mentioned that the main actors are the institutions that are present in the political system of a particular country (Catterberg and Moreno 2005).

The other three variables were constructed in order to test the proposed empirical model as it follows:

- **Political Participation** – signing a petition, joining boycotts, attending peaceful demonstration, joining strikes, and any other act of protest.
- **New sources of information** – usage of internet, email, and mobile as sources for information.
- **Traditional sources of information** – daily newspaper, printed magazines, TV news, radio news, and talk with friends or colleagues as main information sources.

As well, in the analysis there will be used variables that other scholars have found relevant for explaining participation as such, level of education, but also variables such as age and sex.

After a cautious analysis of the data, the results (see Table 1.) of the proposed model (see Annex 1.) show that both new sources of information and traditional sources of information affect the participation level of citizens in Europe. Nevertheless, even if the level of significance of both variables is less than .001, the beta coefficient can be considered as being moderate, though new forms of information can be considered stronger predictors for participation.

Table 1: Participation explanatory model

Explanatory model			*β*	*Sig.*
GENERALIZED TRUST	<---	NEW SOURCES OF INFORMATION	.170	***
POLITICAL TRUST	<---	NEW SOURCES OF INFORMATION	.006	.473
GENERALIZED TRUST	<---	TRADTIONAL SOURCES OF INFORMATION	.281	***
POLITICAL TRUST	<---	TRADTIONAL SOURCES OF INFORMATION	.250	***
PARTICIPATION	<---	NEW SOURCES OF INFORMATION	.214	***
PARTICIPATION	<---	POLITICAL TRUST	.001	.922
PARTICIPATION	<---	GENERALIZED TRUST	.226	***
PARTICIPATION	<---	TRADTIONAL SOURCES OF INFORMATION	.138	***
PARTICIPATION	<---	AGE	.000	.958
PARTICIPATION	<---	SEX	.045	***
PARTICIPATION	<---	EDUCATION	-.099	***

***sig. <0.01

The proposed model also highlighted a difference regarding the effects of generalized and political trust. The model stresses that in the case of political trust the results were not statistical significant. It means that in the proposed model, participation is affected by the level of political trust one has, which comes into contradiction with the stated hypothesis. For generalized trust the results were significant, though the β coefficient is moderate (.226), but slightly better than in the case of new resources information.

In case of the demographic variables that were used in the model, none of the variables were identified as strong predictors for participation. Following this line of argument, education (negative) and sex (positive) influence the level of participation, though the net effects are very low. In this concern, it comes into contradiction with other studies that demonstrated the importance of education and other socio demographic variables for engaging people in different actions of participation. Therefore, summarizing the results, the following equation can be proposed for an explanation of participation within the proposed model:

participation = .226 x generalized trust + .214 x new sources of information + .138 x traditional sources of information + .045 x sex - .099 x education

On the other hand there can also be highlighted other relationships. These results stress the significance of new sources of information in explaining generalized trust, meanwhile in the case of political trust the result is not significant. Traditional sources have a positive effect for generalized and political trust, but better explain generalized trust. For that reason, it can be stated that the types of informational sources people use, shape the level of generalized and political trust within the society. Moreover, besides the fact that new types of resources are more and more part of the daily life, traditional sources of information remain the dominant predictor.

7. Conclusions

The rise of the information society raises a number of new fields for social scientists to focus. It affects every aspect of societies' interactions, and political participation being one of them. Therefore, the access to information can trigger the levels of trust and participation because of the possibility of finding or even disseminating relevant information that is becoming less costly. The amount of resources needed to obtain information is constantly declining. Moreover, alternative points of views are more accessible, meaning that people have the possibility to engage in debates and other forms of participations. New information sources have been identified as having strong effects on participation. Being more explicit, the usage of email, and one's mobile as sources for information do positively influence joining boycotts, signing petitions or joining strikes. Despite stepping into a new era of digitalization in all spheres of the society, it seems that traditional sources of information still play an important role within the democratic processes. Even more, the role of traditional sources of information (daily newspaper, printed magazines, TV news, radio news, and talking with friends or colleagues) play a more significant role for trust in people you meet for the first time and trust in political institutions, meanwhile new forms of information are only significant for generalized trust. Even if at present the information is more accessible, it appears that in order to build political trust it is necessary to go beyond available online sources of information. Moreover it can be assumed that political trust not only relies on traditional sources of information, but can also be affected by side effects such as level of corruption, well-being and many other factors. Nevertheless, political trust did not seem to play an important role in explaining the level of participation, compared to generalized trust that has the strongest effect. It can be assumed that the more trust people will have in strangers consequently it will determine them to engage in various forms of political participation or association. On the other hand, it appears that even if the society trusts or does not trust political institutions, the participation level will be the same. This result can be explained by established democracies, where the level of political culture is high and people will still engage regardless of the trust level.

The empirical results of the relationship between modern and traditional means of information, level of trust and political participation determine to focus on a broader explanatory model. The proposed model suggests the need of diversifying the theoretical approach, including recent development of the technological processes and their interaction with the social and political processes that happen within the society. Therefore, the rapid development of technology determines scholars to adapt to new daily realities and identify new explanatory models for political participation in various societies, though not neglecting the primary sources that cauese people to engage in political actions.

8. References

Aars, Jacob: "Democratic renewal in local government? Top-down strategies for bottom-up involvement". In: Zittel Thomas/ Fichs Dieter (eds.): *Participatory Democracy and Political Participation. Can participatory engineering bring citizens back?* Routledge/ECPR Studies in European Political Science: 2007, pp. 202–222.

Badescu, Gabriel / Uslaner, Eric M.: "Introduction". In: Badescu Gabriel/ Uslaner Eric M. (eds.): *Social Capital and the Transition to Democracy*. Routledge: 2003, pp. 1–9.

Bartkowski, Jerzy: "Social capital in Poland". In: Badescu Gabriel/ Uslaner Eric M. (eds.): *Social Capital and the Transition to Democracy*. Routledge: 2003, pp. 184–199.

Bhuiyan, Serajul I.: "Use of internet in political participation in South Korea". *Asia Pacific Media Educator* 15, 2004, pp. 115–130.

Blind, Peri K.: "Building trust in Government in the Twenty-first Century: Review of Literature and Emerging Issues". 2006, retrieved 20.02.2016, from http://unpan1.un.org/intradoc/groups/public/documents/UN/UNPAN025062.pdf.

Campante, Fellipe, Durante, Ruben and Sobbrio, Francesco: "Politics 2.0: The Multifaceted Effect of Broadband Internet on Political Participation". *NBER Working Paper No. 19029* 2013, retrieved 20.02.2016 from http://www.nber.org/papers/w19029.

Catterberg, Gabriela and Moreno, Alejandro: "The individual bases of political trust: trends in new and established democracies". *Oxford University Press on behalf of the World Association* 18(1) 2005, pp. 31–48.

Cohen, Jean: "Trust, voluntary association and workable democracy: the contemporary American discourse of civil society". Democracy and Trust, Cambridge University Press, 1999, pp. 208–248.

Dahl, Robert A.: "The Past and the Future of Democracy". Revised version of the lecture held at the symposium Politics from the 20th to the 21st Century, CIRCaP, University of Siena, October 14–16, 1999, retrieved 20.02.2016, from http://www.circap.org/uploads/1/8/1/6/18163511/occ_5.pdf.

Dahlgren, Peter: "Social Media and Counter-Democracy: The Contingences of Participation". In: Tambouris Efthimios/ Macintosh Ann/ Sæbø Øystein (eds): *Electronic Particiaption*. Springer: 2012, pp. 1–12.

De Marco, Stefano / Antino, Mirko / Robles Morales, Jose Manuel: "Assessing a Measurement Model for Digital Political Participation: A multidisciplinary Point of View". In: Yannis Charalabidis / Sotirios Koussouris (eds.): *Empowering Open and Collaborative Governance*. Springer: 2012, pp. 61–78.

Fennema, Meindert / Tillie, Jean: "Political participation and political trust in Amsterdam: Civic Communities and ethnic networks". Journal of Ethnic and Migration Studies 25(4), 1999, pp. 703–726.

Finkel, Steven E.: "Reciprocal Effects of Participation and Political Efficacy: A Panel Analysis". *American Journal of Political Science* 29(4), 1985, pp. 891–913.

Freitag, Markus / Bühlmann Marc: "Crafting Trust: The Role of Political Institutions in a Comparative Perspective". *Comparative Political Studies* 42(12), 2009, pp. 1537–1566.

Freitag, Markus: "Bowling the state back in: Political institutions and the creation of social capital". *European Journal of Political Research* 45(1), 2006, pp. 123–152.

Fuchs Dieter: "Participatory, liberal and electronic democracy". In: Zittel Thomas/ Fichs Dieter (eds.): *Participatory Democracy and Political Participation. Can participatory engineering bring citizens back?*, Routledge: 2007, pp. 29–54.

Gainous, Jason / Marlowe, Adam D. / Wegner, Kevin M.: "*Traditional Cleavages or New World: Does Online Social Networking Bridge the Political Participation Divide?*". International Journal of Politics, Culture and Society 26(2), 2013, pp. 145–158.

Glassey, Olivier / Aglione, Camille-Angelo.: "*Getting Teenagers to Participate: A Survey of the youth Council of the City of Lausanne*". In: Tambouris Efthimios/ Macintosh Ann/ Sæbø Øystein (eds): *Electronic Particiaption*. Springer: 2012, pp. 112–119.

Hedtke, Reinhold / Zimenkova, Tatjana: "*Introduction: critical Approaches to Education for Civic and Political Participation*". In: Hedtke Reinhold/ Zimenkova Tatjana (eds.): *Education for Civic and Political Participation*, Routledge: 2013, pp. 1–9.

Karlsson, Martin: "*Democratic Legitimacy and Recruitment Strategies in eParticipation Projects*". In: Yannis Charalabidis / Sotirios Koussouris (eds.): *Empowering Open and Collaborative Governance*. Routledge: 2012, pp. 3–20.

Keeter, Scott / Zukin, Cliff / Andolina, Molly / Jenkins, Krista: "*A guide to the index of civic and political engagement*". College Park, MD: The Center for Information & Research on Civic Learning & Engagement, 2003, retrieved 20.02.2016, from http://pollcats.net/downloads/civichealth.pdf.

Knack, Stephen / Keefer, Philip: "Does social capital have an economic payoff? A cross-country investigation". The Quarterly Journal of Economics 112(4), 1997, pp. 1251–1288.

Kreuter, Frauke / Presser, Stanley / Tourangeu, Roger: "Social Desirability Bias in CATI, IVR, and Web Surveys: The Effects of Mode and Question Sensitivity". Public Opinion Quarterly 72(5), 2008, pp. 847–865.

Letki, Natalia / Evans, Geoffrey: "Endogenizing Social Trust: Democratization in East-Central Europe". *British Journal of Political Science* 35(3), 2005, pp. 515–529.

Letki, Natalia: "Explaining Political Participation in East-Central Europe: Social Capital, Democracy and the Communist Past". Nuffield College Politics Working Paper 2003-W2, *University of Oxford.*

Letki, Natalia: "Investigating the Roots of Civic Morality: Trust, Social Capital, and Institutional Performance". *Springer Science+Business Media* 28(4), 2006, pp. 305–325.

Listhaug, Ola / Ringdal, Kristen: "Civic Morality in Stable, New, and Half-Hearted Democracies". *European Values at the Turn of the Millennium* 7(1), 2004, pp. 341–362.

McClurg, Scott D.: "Social networks and political participation: The role of social interaction in explaining political participation". *Political Research Quarterly* 56(4), 2003, pp. 449–464.

Mische, Ann: "*SPOTLIGHT: Distrust in democracy: complex civic networks and the case of Brazil*". In: Hanagan Michael/ Tilly Chris (eds.): *Contention and Trust in Cities and States*. Springer 2012, pp. 355–359.

Mishler, William and Rose, Richard: "Trust, Distrust and Skepticism: Popular Evaluations of Civil and Political Institutions in Post-Communist Societies". *The Journal of Politics* 59(2), 1997, pp. 418–451.

Mishler, William / Rose, Richard: "What are the origins of political trust?: Testing institutional and cultural theories in post-communist societies". *Comparative Political Studies* 34(1), 2001, pp. 30–62.

Murphy, Kristina: "The role of trust in nurturing compliance: A study of accused tax avoiders". *Law and Human Behavior* 28(2), 2004, pp. 187–209.

Newton, Kenneth "Social and Political Trust in Established Democracies". In: Pippa Norris (ed.): *Critical Citizens: Global Support for Democratic Governance*. Oxford University Press, 1999, pp. 169–186.

Newton, Kenneth: "Trust, Social Capital, Civil Society, and Democracy". *International Political Science Review* 22(2), 2001, pp. 201–214.

Norris, Pippa: "Critical Citizens: Global Support for Democratic Governance". *Oxford University Press* 1999.

Oorschbot, van Wim / Arts, Wil / Gellissen, John: "Social Capital in Europe: Measurement and Regional Distribution of a Multifaceted Phenomenon". *Acta Sociologica* 49(2), 2006, pp. 149–167.

Quandt, Thorsten "What's left of trust in a network society? An evolutionary model and critical discussion of trust and societal communication". *European Journal of Communication* 27(1), 2012, pp. 7–21.

Reeskens, Tim / Hooghe, Marc: "Cross Cultural Measurement Equivalence of Generalized Trust. Evidence from the European Social Survey". *Springer Science+Business Media* 85(3), 2007, pp. 515–532.

Rothstein, Bo / Stolle, Dietlind: "The State and Social Capital: An Institutional Theory of Generalized Trust". *The Journal of Comparative Politics* 40(4), 2008, pp. 441–459.

Schlozman, Kay L. / Verba, Sidney / Brady, Henry E.: "Weapon of the Strong? Participatory Inequality and the Internet". *Cambridge* 8(2), 2010, pp. 487–509.

Schyns, Peggy / Koop, Christel: "Political Distrust and Social capital in Europe and the USA". *Springer Science+Business Media* 96(1), 2010, pp. 145–167.

Tiscornia, Daniela / Fernandez-Barrera, Meritxell: "Knowing the Law as a Prerequisite to Participative eGovernment: The Role of Semantic Technologies". In: Yannis Charalabidis / Sotiris Koussouris (eds.): *Empowering Open and Collaborative Governance*, Springer 2012.

Tyler, Tom R.: "Psychological Perspectives on Legitimacy and Legitimation". *Annual Review of Psychology* 57, 2006, pp. 375–400.

Uslaner Erik M..: "The Moral Foundations of Trust". *Cambridge University Press* 2002.

Uslaner Erik M.: "Trust and governance: how culture and economics constrain the state". In: Zittel Thomas/ Fichs Dieter (eds.): *Participatory Democracy and Political Participation. Can participatory engineering bring citizens back?*. Routledge/ECPR Studies in European Political Science 2007.

Uslaner Erik M.: "Trust but verify: Social Capital and Moral Behavior". *Social Science Information* 38(1), 1999, p. 29–35.

Uslaner, Eric M.: "Democracy and social capital". In:. Mark E. Warren (ed.): "*Democracy and Trust*", Cambridge University Press 1999.

Uslaner, Erik M. / Conley Richard S.: "Civic Engagement and Particularized Trust: The Ties that Bind People to their Ethnic Communities". *American Politics Research* 31(4), 2003, pp. 331–360.

Verba, Sidney / Schlozman, Kay L. / Brady, Henry: "Voice and equality: Civic voluntarism in American politics". *Harvard University Press* 1995.

Woody, Stanley J. / Weare, Christopher: "The Effects of Internet Use on Political Participation: Evidence From an Agency Online Discussion Forum". *Administration & Society* 36(5), 2004, pp. 503–52.

Yamagishi, Toshio / Yamagishi, Midor: "Trust and Commitment in the United States and Japan". *Motivation and Emotion* 18(2), 1994, pp. 129–166.

Appendix: Participation explanatory model and Data

Figure 1: Participation explanatory model

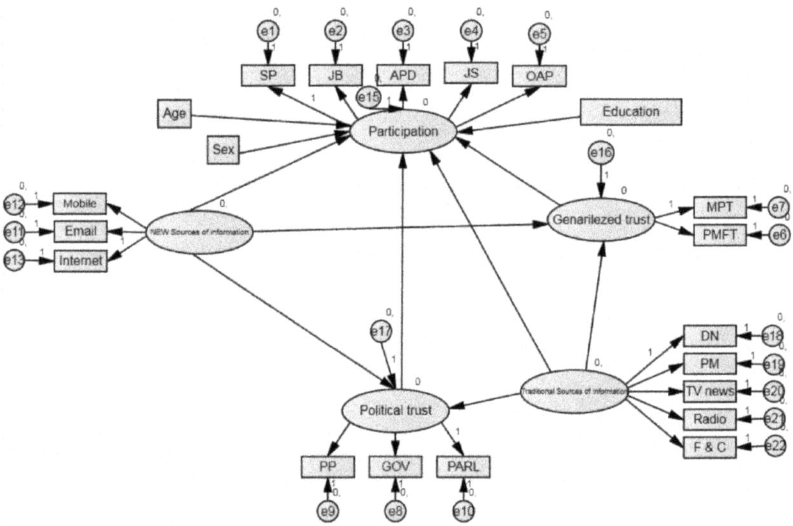

Data 1: WORLD VALUES SURVEY Wave 6 2010–2014 OFFICIAL AGGREGATE v.20141107.

WORLD VALUES SURVEY Wave 6 2010–2014 OFFICIAL AGGREGATE v.20141107. World Values Survey Association (www.worldvaluessurvey.org). Aggregate File Producer: Asep/JDS, Madrid SPAIN.

Matej Makarovič and Mateja Rek
School of Advanced Social Studies in Nova Gorica
Gregorčičeva 19, 5000 Nova Gorica, Slovenia
matej.makarovic, mateja.rek@fuds.si

Experts and the Macro Level Decision-making in the Information Society

Abstract: *In this contribution we explore the role of experts in decision making processes at the macro level in the context of the information society. We apply Niklas Luhmann's social systems theory as our central theoretical and conceptual framework to place the issue in a broader perspective. We analyse why both the decision makers and the general public tend to rely heavily on experts in decision-making processes at the EU and the national level and look for limitations of expectations related to expert knowledge generation, transfers and application in the context of contemporary functionally differentiated information societies. We suggest, that expert supported decision making at the macro level as an expected reaction to increasing societal complexity should be clearly distinguished from the rule by the experts instead of (democratically elected) governments and provide arguments for such a distinction related to the features of the modern and information society.*

Keywords: *experts, information society, policy, EU*

1. Introduction

In this paper we intend to contribute to the understanding of the role of experts in decision making processes at the macro level in the context of the information society. We apply Niklas Luhmann's social systems theory as our central theoretical and conceptual framework to place the issue in a broader perspective. Focusing empirically on the role of experts in the EU decision making processes and the public attitudes towards the role of experts in (political) decision making in general, our purpose is to explain:

1. why both the decision makers and the general public tend to rely heavily on experts in decision-making processes at the EU and the national level;
2. why experts may even be seen as a replacement for the elected politicians;
3. where are the limitations of these expectations in the context of functional differentiation?

First, we will establish a relationship between two key broader concepts essential also for the understanding of the role of experts: the information society and the modern society as distinct but also significantly interrelated phenomena that both contribute to the growth of societal complexity. To the latter, both social systems and individuals as psychical systems have to respond in a variety of ways. Secondly, increasing reliance to experts will be presented as one of the typical responses to the complexity problem. The case of the EU policy making may be an interesting demonstration, how formulation and implementation of policies may be significantly supported by experts. The EU policy making system heavily relies on the comitology, while at the national level the policies are somewhat more strongly determined by party-based political polarisation significantly less present at the European level. Thirdly, a distinction would be required by *experts supported* and *experts determined* policies at the macro level. Although the latter model, where experts would become de facto (and de jure) rulers, could hardly be compatible with a democracy, it has remained a normative ideal from Plato's Republic to the contemporary beliefs of significant parts of the populations of European democracies, both the new and the old ones.

Finally, we intend to explain why the actual expert ruled political system – contrary to the experts supported one – cannot be implemented in the modern society, i.e. not because it contradicts the principles of democracy but because it is incompatible with functional differentiation.

2. Information Society and Modern Society: The Evolution of the Media

According to Luhmann (1995), social systems are auto-poetic self-referential systems that consist solely of communication (instead of action or even people that are placed – as psychical systems – to the necessary environment of the social systems). In general, communication is quite improbable, since every communication attempt is confronted by three major challenges of (1) being understood; (2) reaching the recipient; and (3) being accepted. Social evolution in general, however, results in the development of increasingly improbable systems (Luhmann 2008). This also includes the evolution of the communication media able to reduce the improbability of communication.

Language in the media being is supposed to reduce the improbability of understanding. The challenge of communication reaching the recipient is addressed by the transmission media (e.g. writing, printing, radio, phone, TV, internet, etc.), while the challenge of being accepted is reduced by the symbolically generalised media (e.g. truth, power, law, love, money, etc.). While language may be left aside in our discussion, it may be argued that the information society and the modern society can be seen as descriptions of the evolution of the transmission and the symbolically generalised media respectively.

Both information and modern society should be seen not as (static) situations or stages, which somehow makes the discussions on when the early modernity becomes the late (second) modernity (cf. Beck 1992, Giddens 1991) rather obsolete, but as accelerated *processes* encompassing all aspects of informatisation and modernisation. It would also be conceptually incorrect to understand the information society as a sub-type of the modern society or the late modern society. Instead, both phenomena should be seen as conceptually distinct though clearly strongly interrelated.

What is seen as the information society can be simply described as an accelerated evolution of the transmission media. It should be seen as a process of making the communication reach its recipients significantly less improbable. In other words, with the significantly increased ability of communication to reach its recipients, both social and psychical systems become exposed to enormous amounts of communication able to reach them – in a manner incomparable to any previous situation.

Modernisation, on the other hand, can be described as the evolution of the symbolically generalised media producing the basis for the increasing functional differentiation of the society, i.e. the development of a variety of autonomous, auto-poetic, self-referential functional subsystems, such as science, politics, law, economy, intimate relations, religion, etc. According to Luhmann (1995), these subsystems are not only distinguished by their functions, but also reproduced on the basis of their own particular symbolically generalised media (enabling their specific communications), binary coding (as the only focus of their internal operations), and programmes (as the selection criteria applied for the binary coding processes).

Discussing the interdependencies between the modern and the information aspects of the social systems is beyond the scope of this paper. What needs to be stressed, however, is that both contribute to the new challenges based on the increasing complexity. Functional differentiation can be seen as a response to complexity by increasing specialisation: i.e. a social system only responds to certain (functional, media and code specific) environmental disturbances and ignores the rest. However, increasing functional differentiation generates a new challenge by placing complexity to another level, when different functional subsystems have to be co-ordinated (Makarovič 2001) in terms of contextual interventions, reflectivity and systemic discourses (Willke 1989), or, in other words, when only certain options from the enormous set of options have to be selected.

The enormously increased amount of information reaching the social and psychical systems due to the rapid development of the transmission media, on the other hand, has also significantly increased the complexity with which the (social or psychical) system is confronted. The issue may be understood in terms of Ashby's law or the law of requisite variety claiming that a system should have as many different options available as there are the possible different impulses reaching it from its environment (Ashby 1958). The increased amount of information available through the information society has clearly increased the number of such environmental impulses thus forcing the system to increase its own complexity in order to cope with the increasing variety in its environment generated by the enormous amounts of information. This generates the necessity for both the social systems (Willke 1995) and the individuals (as psychical systems) to develop their reflective abilities (Giddens 1991, Beck 2012, Lash 2010) or – better – place their reflection to a different level from communicative to autonomous and meta-reflectivity (Archer 2012). In other words, they need to be equipped with proper knowledge to be able to select autonomously from the available information and to organise relevant information as knowledge. The problem touches almost everybody: from individual persons (psychical systems) to organisations (social systems); from ordinary people to major policy-makers. In the digital age, there is a vast amount of information potentially available at latter that needs qualifying, authenticating and assessing.

At this point, the information society is inevitably connected to the concepts of the knowledge-based society or the learning society and the learning organisation (Schoen 1983, Hughes and Tight 2012, Davenport 2005). The knowledge required for proper selection may be – at least potentially – more easily reachable but also (1) enormous in terms of comprehending and selecting the relevant aspects and (2) specialised to a very significant extent – at least partially in line with the general lines of specialisation and functional differentiation of the modern social systems. At least due to the latter, the days of the universal wise-persons are clearly over.

These facts imply the increasing need for expertise – for the specialised experts, their collaboration and networks: both on a short-term basis that are dissolved when projects end, and more permanent networks in order to be established for providing a continuous updating of knowledge and an ability to supply fresh knowledge on request.

3. Experts in the EU Policy Making

It was not new to the European integration process when member states chose to ask a group of experts to suggest a possible blueprint for increased integration or policy co-operation. In fact, many studies have shown the Community's integration *modus operandi* to be incremental, based on technical expertise and knowledge, and of an apolitical nature. Recently several legal scholars have studied the role of expert committees in the EC decision-making process, also known as 'comitology', and emphasized the extent of the use of committees in the process (see for instance Gornitzka and Sverdrup 2008, 2011, Rimkute and Haverland 2014, Blom-Hansen 2014). In part the widespread use of expert committees in EU policy-making results from the fact that the European policy-making arena is quite distinct from the national one. Political parties are the prime means by which political power is exercised on the national level. Political parties have the capacity to represent citizens and provide policy choices that demonstrate their ability to govern for the public good. Such national level policy-making is more dependent on the existing political situation, following the decisions taken by voters on national elections. Even if this is the case, the national level policy-making is still faced with similar systemic complexity issues as the transnational one. So, the hierarchical modes of decision-making in national states are also being complemented by horizontal, co-operative forms of policy formulation and implementation. There is a plethora of decision-making arrangements among public, private or civic actors, being experts or citizens, ranging from horizontal and vertical joint decision-systems among public authorities, to policy networks of state and non-state actors, to voluntary or mandated forms of private self-regulation. These differ widely among EU member states as well as the policy area they are framed by.

Unlike in national states at the EU level the equivalent of a national government and parliament is virtually absent even though the concept of the European Parliament is constantly being challenged and developed in the direction to better fill the gap or to say a democratic deficit between the EU institutions and the people of the EU (who don't regard themselves as a community but feel stronger affiliations to the national level). As a result, some traditional political channels, such as political parties, do not play such a significant role as in the national arena. The public deliberation process including broader European public and a truly public debate on European policy-making is lacking as most of the media is nationally oriented, resulting in debates about the EU being held in the national domain. Finally, as it has happened to European governments more generally, European governance has gone through a process of restructuring, which has given rise to a regulatory state instead of an interventionist, redistributive state (Majone 1997). This restructuring process has also contributed to the more frequent consultation of expert committees.

The decision-making system of the EU relies on a plethora of working groups, standardization bodies and committees of experts. Not only citizens but also companies have lamented that EU technocrats working in splendid isolation tend to produce too many rules and questionable legislation. One of the sources of such a state of affairs can be seen in a specialisation of the EU as a political system producing regulatory policy. Knowledge, rather than budget, is the critical resource in regulatory policy-making the Commission utilizes these resources extensively. Regulatory policies aim at efficiency, rather than redistribution. This makes them suitable for discussion and negotiation in expert circles (Radaelli 1999, Radaelli and De Francesco 2007).

Given the specific EU policy environment, the Commission plays a large role in day-to-day policy-making (Birger Skjaerseth and Wettestad 2010). As that institution has only limited staff, it cannot deliver all the specialized expertise required in the policy-making process. Hence, it benefits from listening to and co-operating with expert committees. Likewise, the European Council and the Council of Ministers are also confronted by these institutional restrictions (Hayes-Renshaw and Wallace 2006) and hence also need help from expert committees.

Studies of the Commission offer a picture of Directorates-General (DGs) with different administrative cultures and diverse policy frameworks (Mazey and Richardson 2006). Additionally, policy-making is organized around a certain number of functional policy areas where Directorates of the Commission, Commissioners, and committees of the European Parliament, Council advisory groups, national administrations and interest groups form coalitions competing for power (Richardson 2012).

Moreover, a growing involvement of civic organisations into decision-making processes at the EU level actually implies 'cognitive mobilisation' on one side and, on the other, it means the transformation of these organisations into professional advocacy groups in which grass-root participation plays little role. However, this does not necessarily mean a deviation from democratic principles: in given circumstances it can be an effective response to the processes that are characteristic of the knowledge-based society.

Civic organisations can play an important role as a mediator between politics, business and science and the rest of society and, thus, help in the creation of a new developmental discourse and the making of more carefully considered political decisions.[1] On the other side, new approaches to the sociology of science underline the distinction between 'reliable science' and 'socially robust science' (Nowotny 2003), meaning a new social contract between science and society with the emphases on wider stakeholders and policy deliberation on the long-term implications concerning the implementation of scientific and technological innovations. Civic organisations can also play a very important role as a mediator and 'translator' between scientific expertise and the broader public. It has to be taken into account that: "*Since expertise now has to bring together knowledge that is itself distributed, contextualised and heterogeneous, it cannot arise at one specific site, or out of the views of one scientific discipline or group of highly respected researchers. Rather it must emerge from bringing together the many different 'knowledge dimensions' involved. Its authority depends on the way in which such a collective group is linked, often in a self-organized way*" (Gibbons 1999, p. 11). Put in other words, "[…] *results should be expressed not as single discrete values but as an 'envelope' showing consequences of assumptions. Appraisal methods should be selected for their suitability for extended peer review, including minority opinions*" (Funtowicz et al. 2000, p. 334).

4. Experts and Democracy

Since policy issues faced by the governments have become increasingly complex and since science and technology can severely impact crucial governmental decisions (Snow 2013), the role of providing expertise has become the object of academic interest and passionate debate. Studies have shed light on how knowledge shapes public policy formation owing to the presence of experts and epistemic communities (Richardson 2006, Dunlop 2014), fora of discussion and the policy entrepreneurship of the Commission (Zito 2001), an organization possessing knowledge as its main resource. This seems to bode well in terms of quality of the democratic process. Are we moving then towards better-informed, perhaps even enlightened policy?

Whether inspired by populist prejudices and pressures or purportedly scientific theories and evidence, social policy in market democracies implies a privileging of 'knowledge' or 'expertise': policies are legitimated and rationalized by reference to a claim about their intrinsic validity. It is a crucial source of a policy's legitimacy. Liberal democracy necessarily creates the need for expertise as a basis for social policy since it, first, permits freedom of ideas and competition amongst ideas,

1 Also see Kohler-Koch and Rittberger (2007), who underline the function of civic organisations as agenda-setters and a counterbalance to state and economic actors.

and, second, politically requires government intervention to establish equality of opportunity (though not necessarily of outcomes) amongst citizens by rectifying sources of inequality or by expanding choice. 'Expertise' is a commodity provided by a range of specialists including civil servants, policy analysts and academics, and used by politicians.

Only in such an environment is deliberative democracy,[2] which brings about new actions and a new understanding on the part of civil society, also possible. There is no doubt that in this frame of reference, a certain type of 'elitism' cannot be avoided. It is quite clear that the deliberative democracy model itself presupposes elitist characteristics, such as articulated knowledge, the ability to enter into public dialogue, and well-informed actors.

For both, the EU as a whole and its member states, knowledge-based society and economy remain the goals. This implies that information, innovations, and scientific knowledge transfers, as well as 'knowledge politics' are becoming the main driving factors of (post)modernisation (see Stehr 2003, Stehr 2005). In these terms, long-term strategic decision-making aimed at sustainable development is increasingly important. This sort of decision-making is viable in the model of deliberative democracy; further, this model based on rational discourse and arguments only flourishes in a knowledge-based society. Deliberative politics can only function if actors exist, who:

1. are competent and have the necessary expertise;
2. are oriented towards long-term aims and willing to transcend short-term interests; and
3. are able to formulate the expertise, i.e. policy proposals (including policy-related science) in a broadly understandable but not (too) simplified form of collective narratives (Gibbons 1999). There is no doubt that (some types of) civic organisations are acquiring new significance and a new mission in this context.

Scientific experts are expected to provide 'robust', 'reliable' and 'impartial' advice. However, a variety of environmental and public health scares have highlighted some difficulties faced both by those providing the advice and those who rely upon it. Conflicting scientific claims have meant that answers to scientific questions are largely qualified and uncertain. In many cases, the questions that experts are being asked cannot be clearly answered due to the complexity of the systems being examined (Sanderson 2006).

---

2 For more about this model, see J. Elster, ed. (1998).

Work in the sociology of science has highlighted the uncertain and contestable nature of all science (e.g. Barnes 2013) and involvement in the policy process is seen to exacerbate these problems. Indeed, the contestability of scientific claims has led some authors to argue that science can play little role in policy-making (e.g. Collingridge and Reeve 1986). Funtowicz and Ravetz (1993) are more positive, arguing that in the cases where both system uncertainties and decision stakes are high, science can still be influential. However, they feel that scientific discussion needs to be extended to a wider peer community, incorporating more participants who have an involvement with the issue. Funtowicz and Ravetz (1993), introduce the idea of 'post-normal' science to describe the situation where uncertainty or decision stakes are high, e.g. for many of the situations where scientific input into policy-making is needed. As quality of science is maintained through the process of peer-review, the quality of post-normal science should be maintained through review by extended peer communities, which include a wide range of stakeholders to the issue.

A further possible approach to generating advice on scientific issues is through a much greater involvement of the public. Many different methods for engaging the public have been proposed including: deliberative polling; standing panels; focus groups; citizens' juries; consensus conferences; stakeholder dialogues; and internet forums and dialogues. A greater use of the public is seen by its advocates as not only potentially beneficial in terms of improving the quality and accountability of advice that may be generated, but also has a way of generally improving the political process. This has stimulated the greater involvement of society in the debate surrounding scientific and technological issues, and has raised the issue of ethics to the core of much of this debate. On the other hand, these developments have been matched with heightened concern over the utility and validity of scientific advice and the acceptance of consequences of its use by general public, for instance, is the general public aware of potential benefits and risk involved in the development and the use of nanotechnology? Are these acceptable? Do people want to live in a "Nano society"? Do they have a choice?

Political and administrative decision-making may also become more open and transparent on all levels. This trend answers to the demands of citizens and interest groups for more information on the grounds and procedures of decision-making and leads to increased accountability of the involved actors[3].

3 To an increasing extent citizens can find out via the Internet, which interest groups (for example employers' associations, non-governmental organisations) have been consulted in the decision-making process and which arguments have been brought forward.

Policy-making with respect to science and technology is also becoming more inclusive. This means that previously closed policy-circles are breaking up and new actors, mostly from civil society, are becoming involved. Also the basis of policy relevant knowledge is getting more inclusive. Increasingly, the monopoly of scientific experts for the supply of expertise is questioned. The specific knowledge of practitioners, consumers, users or patients is increasingly gaining importance for evaluating policy options, i.e. expertise is "democratised". Hence, the notion of "science" is changing due to the fact that a scientific research process will have to integrate other people than only scientists. Science itself is held accountable by society for its choice of research topics and comes under pressure to serve societal (or economic) demands. The agenda of research – as can be seen for example in the case of foresight – is being formulated more and more according to societal demand, and less by independent "basic research". As a consequence, there is a growing demand for policy tools to raise "scientific literacy" of the public (Banthien et. al 2003).

As we have claimed, the common cause of these trends lies in the increasing complexity of the societal and scientific frames. Society is increasingly interlinked and diverse, both horizontally (with respect to the functional differentiation plurality of knowledge, interests and values) and vertically (from micro- to transnational levels). Scientific and technological progress raises new issues, both concerning the increased awareness of ignorance that – seemingly paradoxically – has arisen from increased scientific knowledge, and concerning the normative evaluation of new options for action. In addition, policy-making has to face global challenges (such as climate change, droughts and floods, population growth, poverty, terrorism), with the sustainability of future developments emerging as the pivotal yardstick that is, at the same time, highly abstract and demanding.

5. Experts or Democracy

Even when experts are linked to civic organisations their increased role is not necessarily compatible with the usual understanding of the concept of (representative) democracy. The dilemma was nicely formulated by Saurugger (2006): "[...] *we have to question the automatic link we draw between civil society representation and democracy...The better informed and organized a group, the greater its chances are to gain access to the European institutions. Expertise and perceived efficiency are central access goods for civil society...However, this may lead to an expertise-representation gap...: the better structured and organized group is, and the more it is therefore able to offer necessary expertise, the less its members feel represented*" (Saurugger 2006, p. 261).

The problem of politics and representative democracy – and in this specific sense we can talk about an aspect of (output) democratic deficit – is that the framework of its political acting and decision-making is narrow, bound by the short-term horizon of one or two electoral mandates. Functional subsystems are only able to operate in terms of their binary coding, i.e. government/opposition in the case of the political subsystem (Luhmann 1990), which implies that decision making is indistinguishable from the electoral concerns.

In practical terms this was expressed by J. C. Juncker, President of the European Commission, when explaining why the EU had not implemented the measures and reforms of the Lisbon agenda: 'We all know what we need to do, but we don't know how to win elections after we do it'. If politicians have to carry out certain measures and reforms that would have a positive effect in the medium or long term, with no effect evident in the short term, or which would face opposition during the four-year mandate, then politics tends to avoid such measures and reforms. Therefore, a sufficient consensual critical mass of strategic thinking simply does not exist. Proceeding from the 'trends foresight', experts may be called upon to contribute to this critical mass and to enrich political discourse with elements of cognitive complexity, reflection and deliberative decision-making. However, should experts be seen as the key support for the decision making by democratically elected politicians or even as the key decision makers in the future?

Given the extent of involvement of experts in policy-making processes it seems legitimate to ask the following questions: How do Europeans feel about experts being involved in policy-making? Do they have confidence in them? As we are not aware of any research that would offer us a direct answer to these questions we aimed to answer these simple questions by applying data from the World Values Survey (WVS) and European Values Survey (EVS). In both cross-national surveys we find data on the Europeans attitude towards "having experts, not government, make decisions according to what they think is best for the country[4]"

The survey question implies not only the support role of the experts in the decision making processes but also the experts acting *instead of* (a democratically elected) government. This understanding can also be observed in the answers to the survey questions, since the correlation between the support for democracy

4 The survey question: I'm going to describe various types of political systems and ask what you think about each as a way of governing this country. For each one, would you say it is a very good, fairly good, fairly bad or very bad way of governing this country? Having experts, not government, make decisions according to what they think is best for the country. Possible answers: 1.) Very good 2.) Fairly good 3.) Fairly bad 4.) Very bad.

and support for the experts making decisions was moderately negative with the Spearman correlation coefficients of -0.12 for the 1999–2001 wave of the European Values Survey and -0.09 for the 2008–2010 wave (significance P equalled 0.000 for both waves).

Table 1: Having experts, not government, make decisions – data from European Values Survey, 2008

	Very good	Fairly good	Fairly bad	Very bad
Austria	14.1	44.1	25.3	16.4
Belgium	11	49.7	30.3	9
Bulgaria	37.2	52.4	9.3	1
Cyprus	9	16.1	21.5	53.4
Czech Republic	19	46.4	22.3	12.4
Denmark	7.6	20.9	34.7	36.8
Estonia	14.3	48.4	29.7	7.5
Finland	6	43.6	36.9	13.4
France	8.1	43.5	30.9	17.5
Germany	16.3	47.1	22.8	13.8
Greece	4.1	14	23.3	58.6
Hungary	33.7	50.7	10.4	5.2
Ireland	13.3	35.4	23.9	27.4
Latvia	16.8	53.6	24.9	4.7
Lithuania	10	42.6	38	9.5
Luxembourg	11.3	35.4	33.9	19.4
Malta	12.4	33.5	29.5	24.6
Netherlands	6.5	44.2	35.5	13.8
Northern Ireland	9.1	42.5	30.2	18.2
Poland	21.2	56.2	17.7	5
Portugal	12.1	51.5	31	5.3
Romania	32.6	41.4	19	7
Slovak Republic	22.7	62	11.9	3.3
Slovenia	23.2	58.1	16.4	2.3
Spain	14.2	48.3	22.7	14.9
Switzerland	6.1	32.6	38.3	23
Total	17.9	43.8	25	13.3

Source: European Values Survey, 2008 (fourth wave).

From Table 1 we can see that a majority of Europeans (with the exception of those from Cyprus, Denmark, Finland, Greece, Luxemburg, Malta and Switzerland) have a positive attitude towards having experts, not government making decisions in policy-making. Similar argument can be drawn on the basis of the data gathered in the World Values Survey (in the year 2000 and 2005), where the same question has been asked.

The reason why we are presenting the data from the three waves of the European Social Survey (ESS) and World Values Survey (WVS) is to allow for longitudinal dimension of analyses, which is presented in Figure 1. It is interesting to note that in a framework of the last 10 years, a lot of 'traditional subjects of governance' like political parties, parliaments and governments have been losing support – the trust, confidence and positive attitudes of Europeans have been decreasing. According to the Eurobarometer surveys, the per cent of people trusting their national governments has declined from 34 to 27 per cent from 2004 till 2014, while the per cent of those that trust their national parliaments declined from 38 to 28 per cent during the last ten years (Standard Eurobarometer 81, Spring 2014). Nevertheless, when it comes to experts making policy decisions, the positive attitude has remained stable for the past 10 years as indicated in Figure 1. Traditionally, support for the rule of the experts as an alternative to (elected) government has been higher in new democracies (i.e. those that were not democratic before the end of the 1980s) than in the old ones, which might be attributed both to the better functioning democracies in the old EU member states as well as to the more consolidated political culture supporting representative democracy. However, from the 1999–2001 survey wave till the 2008–2010 one, the support for the rule by the experts has even slightly increased in the old democracies and remained stable in the new ones.

Figure 1: Longitudinal analyses of the attitudes of Europeans towards having experts, not government, make decisions – comparison of 3 waves (WWS 2000, WWS 2005, EVS 2008)

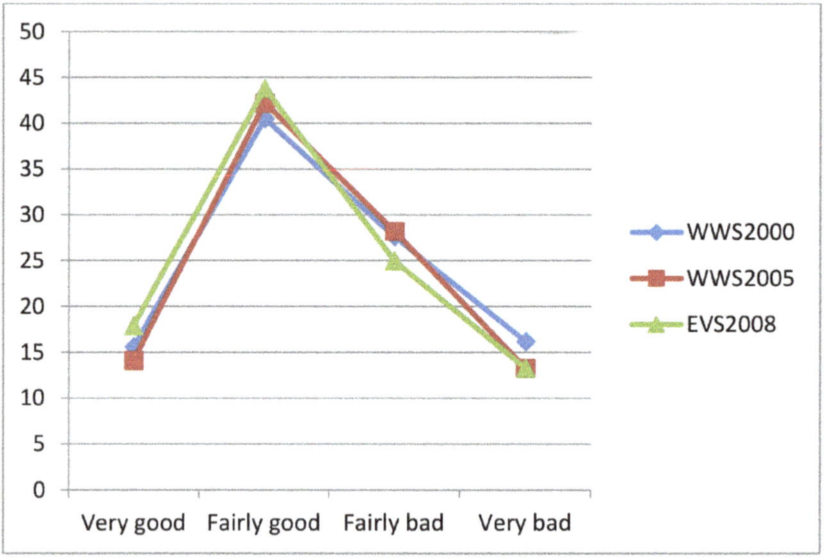

** Compared are total values as presented in Tables 3, 4 and 5*

Source: European Values Survey 2008, World Values Survey 1999/2000, 2005.

In a knowledge based society we learned to value knowledge, skills and competence, which on macro level, have become the key resources and initiators of economic and social change. Production and the application of knowledge have become associated with progress, greater security, the higher quality of people's lives, and the prosperity of societies. The use of knowledge is associated with the fulfilment of social needs and the provision of higher living standards. The new social opportunities are created in the way that »rational knowledge«, which is created within one system (science/research/development), traverses the borders of various social sub-systems (Bell 1973). Common sense explanations are often being replaced by scientific definitions and definitions of reality. In all areas of social and personal life people are turning to knowledgeable individuals or to say individuals who poses specialized knowledge in search for answers and solutions to their own problems and challenges. As in other areas of social life, knowledge and experts are also becoming the most important resources for policy decision-making and collective action (Willke 2007).

The question of positive attitudes towards experts raises interesting issues about their democratic potential as well as possible threats to democratic decision-making processes. In an era when positive attitudes and confidence in traditional political channels are in decline and when an even greater number of Europeans are expressing dissatisfaction with the way national democracies are developing, the stability of positive attitudes towards experts (not democratically elected governments) making decisions and relatively high levels of confidence in expert based decision making poses new dilemmas about the future development of governance and democracy in the EU.

6. Conclusion

Basing key decisions in politics – and other social subsystems – on knowledge and expertise is a necessary reaction to the increased complexity related both to the functional differentiation of modern society and the massive inflows of information as the key feature of the information society. As argued by Helmut Willke: "Many decisions simply do not need moral references anymore because they now can be grounded on evidence-based expertise" (Willke 2009, p. 32). Ideological divisions and struggles, so typical for a classical political arena may thus be overcome to a significant extent by technocratic expert-based decision making. Where they have remained – and they have clearly done so in many cases – they have been preserved and reproduced more as a remnant of the past than as a productive perspective for the future policies.

However, expert supported decision making at the macro level as an expected reaction to increasing societal complexity should be clearly distinguished from the rule by the experts *instead of* (democratically elected) governments. While the former is necessary the letter would be hardly problematic. Our objections against the rule by the experts as a replacement for elected politicians are not based so much on our value-normative judgments favouring (representative) democracy but simply on the following arguments related to the features of the modern and information society:

1. Perhaps, knowledge and expertise may be distinguished from value-normative judgments and interests for the analytical and heuristic purposes or in terms of Weberian ideal types but not in real-life situations. Even the right to be recognised as an expert for the decision making purposes may be related to power-based judgements. This issue is related to the problem of translation between different symbolically generalised media and different codes applied by different social subsystems. The operations of the political subsystem are

based on the media of power and the government-opposition distinction. The scientific media of truth thus cannot be simply 'translated' into power in a way that truth would just become power or even replace it. While science is based on the perpetual quest for truth, which is never completed, the political subsystem is required to make final decisions with real consequences in real time, usually in the situations, where truth (in terms of understanding the factual complexity of the system and all the possible consequences of political interventions) is only available to a very limited extend – even when political decision making is highly supported by the expertise. In other words, when entering the political subsystem as a decision maker, an expert inevitably ceases to be an expert and becomes a politician (even when manipulating the masses by pretending to be something else).

2. Modern society is highly differentiated in functional terms, which implies that certain functions can only be performed by the subsystems specialised in this regard. Science as a subsystem cannot take the role of the political subsystem, without totally denying the principle of functional differentiation. Such dedifferentiation attempts would not fit the challenges of complexity, i.e. 'scientific politics' or 'political science' would be unable to cope with the challenges of complexity for both the political and the scientific subsystem. Although there has been a variety of attempts in this regard in theory and practice from Plato's *Republic* to 'scientific socialism' (cf. Makarovič 2009), they did not prove to be successful – not even in social systems of much lower levels of complexity. Moreover, they have been essentially *anti-political* (see Ryan 2012) regarding Plato's Republic), denying any necessary functional autonomy both to the political system and – more implicitly – to the scientific one as well.
3. Expertise in the modern and information society is highly specialised, mostly also in lines with its functional differentiation. Experts in modern-information societies thus cannot be seen as persons of universal wisdom and impartiality in terms of Plato's Guardians but mostly as specialists able to observe and represent only certain aspects of their social and other environments.

Decision-making processes have become extremely complex in the modern and information societies. While they cannot depend neither solely on elected politicians, nor qualified experts, they can only be taken in a complex interplay including political power and scientific expertise in terms of knowledge through the processes of contextual intervention, increasing reflectivity of psychical and social systems and systemic discourses (cf. Willke 1995). Only such combinations may represent a viable way of addressing the societal complexity based on the principles of the information and modern society.

7. References

Archer, Margaret S.: *The Reflexive Imperative in Late Modernity.* Cambridge University Press: Cambridge. 2012.

Ashby, W. Ross. "Requisite Variety and its Implications for the Control of Complex Systems." *Cybernetica* 1(2), 1958, pp. 83–99.

Banthien, Henning / Jasper, Michael / Renner, Andreas: *Governance of the European Reserach Area: The Role of Civil Society. Final Report.* Retrieved 19. 1. 2015 from: http://ec.europa.eu/research/science-society/document_library/pdf_06/final_report_study-20-october-2003_en.pdf

Barnes, Barry: *Scientific Knowledge and Sociological Theory.* Routledge & Kegan Paul: London and Boston. 2013.

Beck, Ulrich: *Risk Society: Towards a New Modernity.* SAGE Publications. 1992.

Beck, Ulrich: *Twenty Observations on a World in Turmoil.* Polity Press: Cambridge. 2012.

Bell, Daniel. 1973. The Coming of Post-Industrial Society. A Venture in Social Forecasting. New York: Basic Books.

Birger Skjaerseth, Jon / Wettestad, Jorgen: "Making the EU Emissions Reading Systems: The European Commission as an Enterpreneur Epistemic Leader". *Global Environmental Change* 20(2) 2010, pp. 314–321.

Blom-Hansen, Jens 2014: "Comitology Choices in the EU. Legislative Process: Contested or Consensual Decisions?" *Public Administration*, 92(1) 2014, pp. 55–70.

Collingridge, David / Reeve, Colin: *Science Speaks to Power: the Role of Experts in Policy Making.* Frances Pinte: London. 1986.

Davenport, Thomas H. *Thinking for a Living. How to Get Better Performance and Resultr from Knowledge Workers.* Harvard Business Press. 2005.

Dunlop, A. Clair: "The Possible Experts: How Epistemic Communities Negotiete Barriers to Knowledge use in Ecosystems Setvices Policy". *Government and Policy*, 32(2) 2014, pp. 208–228.

Elster, John 1998: *Deliberative Democracy.* Cambridge University Press: Cambridge. 1998.

Eurobarometer Surveys. *Standard Eurobarometer 81*, Spring Wave 2014, retrieved 19. 1. 2015 from http://ec.europa.eu/public_opinion/archives/eb_arch_en.htm.

European Values Study, retrieved 19. 1. 2015 from http://www.europeanvalues study.eu/.

Functowicz, Silvio O. / Ravetz, Jerome R.: "Science for the Post-Normal Age". *Futures*, 25(7) 1993, pp. 739–775.

Funtowicz, Silvio et al.: "Science and Governance in the European Union: a Contribution to the Debate". *Science and Public Policy*, 27(5) 2000, pp. 327–336.

Gibbons, Michael: "Science's New Social Contract with Society". *Nature*, 2 1999, pp. 11–18.

Giddens, Anthony: *Modernity and Post-Modernity: Self and Society in the Late Modern Age*. Polity Press: Cambridge. 1991.

Gornitzka, Asa and Sverdrup, Ulf: "Who are the Experts? The Information Basis of EU Decision-Making", *ARENA Working Papers*, Center for European Studies, University of Oslo. 2008.

Gornitzka, Asa and Sverdrup, Ulf: "Access of Experts: Information and EU Decision-making. West European Politics", 3(1) 2011, pp. 48–70.

Hayes-Renshaw, Fiona / Wallace, Helen: *The Council of Ministers*. Palgrave MacMillan: Basingstoke. 2006.

Hughes, Christina /Tight, Malcom: "The Myth of the Learning Society Revisited". In: Kerry, Trevor (ed.): *International Perspectives on Higher Education: Challenging Values and Practices*. Bloomsbury Publishing. 2012, pp. 157–173.

Keough, Kevin: "Science Advice for Government Effectiveness: the Canadian Approach". *IPTS Report*, 45 2000.

King, Desmond / Hansen, Ronald: "Experts at Work: State Authonomy, Social Learning and Eugenic Sterilization in 1930 Britain". *British Journal of Political Science*, 29(1) 1999, pp. 77–107.

Kohler-Koch, Beate / Rittberger, Berthold: *Debating the Democratic Legitimacy of the European Union*. Lanham: Rowman & Littlefield Publishers. 2007.

Lash, Scott: *Intensive Culture: Social Theory, Religion and Contemporary Capitalism*. Sage Publications Ltd. 2010.

Luhmann, Niklas: *Political theory in the welfare state*. Walter de Gruyter. 1990.

Luhmann, Niklas: *Social Systems*. Stanford University Press. 1995.

Luhmann, Niklas: *Risk. A Socological Theory*. Transaction Publishers. 2008.

Majone, Giandomenico: "From the Positive to the Regulatory State: Causes and Consequences of Changes in the Mode of Governance" *Journal of Public Policy*, 17(2) 1997, pp. 139–167.

Makarovič, Matej: *Usmerjanje sodobnih družb*. Ljubljana: Znanstveno in publicistično središče. 2001.

Makarovič, Matej: *A failure in handling societal complexity : communist regimes from the perspective of the social systems theory*. Variant: Moscow. 2009.

Mazey, Sonia / Richardson, Jeremy: "Interest Groups and EU Policy-making". In: Richardson, Jeremy (ed.): *European Union. Power and Policy-making*. Third edition. Routledge. 2006.

Nowotny, Helga: "Re-Thinking Science: From Reliable Knowledge to Socially Robust Knowledge". In: Lepenies, Wolf (ed.): *Entangled Histories and Negotiated Universals. Centres and peripheries in a Changing World*. Campus Verlag. 2003.

Radaelli, Claudio M.: "The Public Policy of European Union: Whither Politics of Expertise?" *Journal of European Public Policy*, 6(5) 1999, pp. 757–74.

Radaelli, M. Claudio / Francesco De Fabrizio: Regulatory Quality in Europe: Concepts, Measures and policy Processes. Manchester University Press. 2007.

Richardson, Jeremy: *European Union. Power and Policy-making*. Third edition. Routledge. 2006.

Richardson, Jeremy: Constructing a Policy-Making State? Policy Dynamics in the EU. Oxford University Press. 2012.

Rimkute, Dovile and Haverland, Markus: "How does the European Commission use Scientific Expertise? Results from a Survey of Scientific Members of the Commission's Expert Committees". *Comparative European Politics*, retrieved 16. 5. 2014 from: http://www.palgrave-journals.com/cep/journal/vaop/ncurrent/abs/cep201332a.html.

Ryan, Alan: *On politics: A history of political thought from Herodotus to the present*. London: Penguin Books. 2012.

Sanderson, Ian: "Complexity, Practical Rationality and Evidence-based Policy Making". *Policy & Politics*, 34(1) 2006, pp. 115–132.

Saurugger, Sabine: "The Professionaliation of Interest Representation: a Legitimacy Problem for Civil Society in the EU?" In: Smismans, Stijn (ed.): *Civil Society and Legitimate European Governance*, Edward Elgar, 2006, pp. 260–276.

Schon, Donald A. (1983): *The Reflective Practitioner. How Professionals Think in Action*. Basic Books. 1983

Snow, Charles: Science and Government. Harvard University Press. 2013.

Stehr, Nico: The Social and Political Control of Knowledge in Modern Societies. *International Social Science Journal*, 55(178) 2003, pp. 643–655.

Stehr, Nico: *Knowledge Politics: Governing the Consequences of Science and Technology*. Paradigm Publishers. 2005.

Willke, Helmut. *Systemtheorie III: Steuerungstheorie: Grundzüge einer Theorie der Steuerung komplexer Sozialsysteme*. Lucius & Lucius. 1995.

Willke, Helmut: *Systemtheorie Entwickelter Geselschaften*. Beltz Juventa. 1989.

Willke, Helmut. 2007. Smart Governance: Governing the Global Knowledge Society. Frankfurt and New York: Campus Verlag.

Willke, Helmut: *Governance in a Disenchanted World: The End of Moral Society*. Edgar Elgar Press. 2009.

World Values Survey, retrieved 19. 1. 2015 from: http://www.worldvaluessurvey.org/.

Zito, R. Anthony: "Epistemic Communities, Collective Enterpreneurship and European Integration". *Journal of European Public Policy*, 8(4) 2001, pp. 585–603.

Janja Mikulan Kildi
School of Advanced Social Studies in Nova Gorica
Gregorčičeva 19, 5000 Nova Gorica, Slovenia
janja.mikulan@fuds.si

Crisis Mapping for More Inclusive Humanitarian Information Management in the ICT Era

Abstract: *Structural changes such as development of Information and Communications Technology (ICT) demand more inclusive, diverse and bottom-up forms of Humanitarian Information Management (HIM) and a reflection on opportunities and constraints of such changes for the humanitarian sector and involved participants. The article addresses the issue of the mapping and final results enabled by ICTs and crowdsourcing that have a great potential to address two significant issues related to the management of humanitarian information. These are limited geographic information as well as limited information and knowledge about emergency relief actors in the field, and weak involvement of (local) communities and other relevant actors for crisis preparedness and response. The article presents the case study of EUROSHA volunteers in Kenya who were aiming at crisis preparedness by increasing the level of information of local and international (humanitarian) actors through open-source tools and ICTs. EUROSHA volunteers managed to involve different relevant actors into HIM, have increased the mapping capacities of several actors, and have designed or upgraded the open source maps of four rural areas in the mid-West part of Kenya. However the intense participant observation also exposed a need to monitor and evaluate the impacts of such maps and mitigate the challenges related to (crisis) mapping.*

Keywords: *Humanitarian Information Management (HIM), Information and Communications Technology (ICT), inclusiveness, crisis mapping, crowdsourcing, EUROSHA, Kenya.*

1. Introduction

The speed, the width and the depth of technology development has a significant impact on structures in developed as well as developing countries. Therefore it is not surprising that humanitarian aid agencies and other relevant actors involved in the humanitarian sector are questioning how these developments will change their functioning. This is clear from several studies and reports issued on this topic in recent years such as Disaster Relief 2.0 (2011), NetHope Case Study (2011), Crisis Mapping in Action: How Open Source Software and Global Volunteer

Networks Are Changing the World One Map at a Time (2012), Humanitarianism in the network age (2013), etc.

Such structural changes that are quickly taking place will certainly offer many new opportunities but also many uncertainties regarding people's behaviour in emergency situations. Among those are the changing roles of actors and stakeholders in crisis preparedness and response, interactions between organizations and individuals, the distribution of power, the adaptation in the delivery of humanitarian assistance in the context of more informed and connected individuals, etc. Some conclusions of the above-mentioned studies suggest a fundamental shift in power from capitals and headquarters to the people aid agencies aim to assist. These definitely present the need for modification of HIM since it demands more inclusive, diverse and bottom-up forms of decision-making. Therefore traditional systems are facing the reality where information can be generated by anyone and shared with anyone. The professionalization and commercialization of the OpenStreetMap (OSM) project that will be presented later on, illustrates the new institutional regime that governs cartography within the new era. Although state agencies remain the references for cartographic data, the increasing participation of citizens and the private sector signals the replacement of the state as the main authoritative reference for the collection and diffusion of geographic information (Leszczynski in Caquard 2014). A more inclusive HIM system is a recognised objective for the humanitarian community, but there is no established scope of application yet. "HIM community is still testing and piloting the integration of Volunteer and Technical Communities (V&TC) in preparedness and crisis response." (OCHA 2013). Stephenson and Schnitzer (2009) argue that polycentricity maybe a more suitable aspiration for participants in the international relief system than are calls for new and more severe forms of monocentric coordination, but its conditions too maybe difficult for the relevant participants to realize. The article focuses on opportunities, achievements and constraints of more inclusive HIM, more precisely on volunteers providing the geographical information from the bottom-up approach, making it available on the open-source platforms, cooperating with actors relevant for crises preparedness and response, and training them on mapping techniques.

The main objective in the first part of the article is to illustrate some of the main ideas regarding HIM in the ICT era and, opportunities of crowdsourcing and crisis mapping through a critical literature review. While the second part presents the case study of the EUROSHA project in which volunteers were responsible for introducing a new scope of application of HIM to the field actors. In the framework of the case study we have conducted an intensive participant observation in Kenya during which we found ourselves closely involved – participating in and observing – the everyday lives of a particular set of actors or informants. Secondly, as stressed by Clifford (1992) and Stewart (1998) participant observation obliges the fieldworker to take into account all aspects of a particular community of people. It thus enables us to synthesise disparate and apparently disconnected observations into an integrated whole, and thereby to arrive at a holistic interpretation or construct of the social or cultural form under study.

2. Missing Opportunities of Humanitarian Information Management in the ICT Era?

Management of information in the context of a humanitarian crisis caused by natural or man-made disasters present a specific challenge. From closed or destroyed infrastructure relevant for providing the information such as government offices, NGOs, etc., broken phone and internet connections to inaccessibility of essential individuals responsible for information management, etc. Besides damage caused by natural disaster or conflict, developing countries are often facing poor baseline data and limited access to the relevant information due to the non-established government systems or agreements allowing disaster management organizations or external humanitarian responders to access this information. As stressed by OCHA (2013) some countries simply do not have resources or the statistics capacity and therefore lack basic data that would improve response, such as census information, community capacities and resources. Therefore decision makers and responders often dispose with very limited and even inaccurate information, and consequently respond in an inefficient and ineffective way.

Further the new models of managing humanitarian information must adapt to changing realities where more actors are involved in crisis and more people have access to ICTs. On one hand we can observe that trends in information technology are also reflected in the trends in HIM, including hosting and distributing geospatial data sets. As stressed by Payne et al. (2012, p. 132) data warehouses remain an important part of emergency response because preserving and reusing geospatial data can save time and money. The current landscape of portal management is characterized by new platform architectures, data sources, and data pathways. In light of these developments, repository managers are evaluating how they will deploy, manage, and develop their portals and associated activities, taking into account the new expectations of affected populations, users, contributors, and information management officers and the wider humanitarian community. However various Real-Time Evaluations of different crise such as the cases of Pakistan, Somalia and Ethiopia disclosed that the lack of good quality information and information sharing has been a major barrier for partnering and responding in a successful manner (OCHA 2013). Moreover the humanitarian sector does not take advantage of the opportunities ICTs are offering. Nethope, a collaboration of 43 humanitarian organizations that strives to make the best use of technology, both at the office and in the field, is stressing that humanitarian information remains quarantined and in silos, and affected communities, while consulted, are not involved in a two-way process that verifies and validates decisions (Nethope 2015).

Despite the official recognition that communication with affected communities is a crucial step towards more inclusive humanitarian assistance, evaluations (such as those on Pakistan, Philippines and Mozambique) are constantly revealing the insufficiency of such communication and the fact that affected people are not placed at the centre of decision-making. According to the era we live in, ICTs are definitely offering numerous opportunities for intensifying these relations. While governments and humanitarian organizations are recognizing the need to take advantage of new data sources, there is still a tendency for people removed from a crisis to decide what is best for the people living through that crisis. Consequently they are missing lots of opportunities for improving the response (IRGC 2009; Disaster Relief 2.0 2011; OCHA 2013).

Three particular features of data in the modern age offer significant opportunity for better humanitarian decision-making: a) big data: massive volume of data stored by the governments, the private sector and individuals; b) rich data through Geographical Information Systems (GIS) which combine hardware and software used for the storage, retrieval, mapping and analysis of geographic data; c) open data: unlocking the potential value of big and rich data requires the data to be accessible and free or "open". The standard definition is that data is only open when anyone is free to use, reuse, and redistribute it. Moreover data must also be technically open, i.e. available in standard, editable file formats, easy to access and download, searchable and catalogued. Ideally, data should be well documented to make it easier to use (OCHA 2013).

In the ICT era and so called Web 2.0 era (read plus write era) of the internet that is differentiated by participation as opposed to broadcast and consumption various techniques have developed in order to gather, organize and analyse the information, in particular crowdsourcing, crowdseeding and mobile cash. Since the case study is mostly related to crowdsourcing, we are presenting some of the main ideas with specific respect to mapping.

2.1 Crowdsourcing for the production of geographic knowledge

"Crowdsourcing" is a quite recent concept that incorporates many different practices; therefore there are numerous definitions of it. The paper is applying the definition of Estellés-Arolas and González-Ladrón-de-Guevara (2012) that managed to analyze existing definitions, extracted common elements and presented one of the most comprehensive definitions. They define crowdsourcing as a "type of participative online activity in which an individual, an institution, a non-profit organization, or company proposes to a group of individuals of varying knowledge, heterogeneity, and number, via a flexible open call, the voluntary undertaking of a task. The undertaking of the task, of variable complexity and modularity, and in which the crowd should participate bringing their work, money, knowledge and/or experience, always entails mutual benefit. The user will receive the satisfaction of a given type of need, be it economic, social recognition, self-esteem, or the development of individual skills, while the crowdsourcer will obtain and utilize to their advantage that what the user has brought to the venture, whose form will depend on the type of activity undertaken" (Estellés-Arolas and González-Ladrón-de-Guevara 2012, pp. 9–10). One of the significant tasks that have to be solved are those related to disasters. People from around the world can act as an emergency responder by fulfilling simple tasks in the field (for example getting information directly from the communities) or online (such as geo-tagging) which in the past have proven to be a valuable support to humanitarian aid agencies. The crowd can also just act as a sensor or social computer, where their real-time online reports in social media can contain useful information during a crisis. Therefore ICTs are enabling the significant involvement of new actors on local, national, regional and global levels. International organizations such as the United Nations and the World Bank are increasingly joining crowdsourcing projects and seek support from upcoming Volunteer and Technical Communities (V&TC) performing crowdsourcing. There are many advantages over traditional approaches fostering these movements such as rapid, cost-efficient and relatively accurate and visual analysis of complex situations demanding quick response (Hörler 2014, 1). At such increased involvement of individuals voluntarily contributing the data, questions are being raised regarding their motives and sustainability. Some authors (Grira et al. 2009; Wilson 2011b in Caquard 2014) explain this by referring to different personal motives such as idealism or local needs. Dodge and Kitchin (2012, p. 20) stress that the apparent willingness of many people to participate for 'free' in crowdsourcing projects is undoubtedly based on the fact that they provide genuinely effective platforms to connect socially, communicate meaningfully, and contribute collectively. However there are ongoing debates around the worth of these connections and communications, with critics arguing that they are superficial, lacking depth and obligatory commitment (Carr in Dodge and Kitchin 2012, p. 20). This is definitely related to the issue of sustainability we have raised and upgraded with concerns over the unpredictability of crowds, their narrow demographic profile (digital elites), the quality and consistency of content and metadata created across diversely skilled and motivated individuals, degrees of reliability, a sense of trustworthiness, etc. As it will be presented further on, most of these challenges of crowdsourcing were clearly visible in the EUROSHA project as well.

"Crowdsourcing is radically transforming the acts of making and using maps, fusing what were traditionally two separate practices, and altering fundamentally the political economy of the cartographic industry" (Dodge and Kitchin 2012, p. 34). There are various theoretical positions explaining the attraction, power and problems of crowdsourcing and mapping such as Ritzer's notion of 'prosumer societies', Carr's (2007) critique of the 'ignorance of crowds' and Goodchild's (2007) conception of volunteerist "citizen scientists". In the context of the EUROSHA, the latter in particularly interesting. Goodchild (2007) argues that one of the key features of new forms of geographical knowledge creation (neogeography) premised on Web 2.0 is "volunteered geographic information" (VGI), effectively geocrowdsourcing. He observes that contemporary crowdsourcing can be seen as following longer established trends in "citizen science" which are "communities or networks of citizens who act as observers in some domain of science". People provide empirical observations and detailed measurements of phenomena for free, which are, crucially, of real value as data points for "proper" science. Citizen scientists are particularly useful in observational terms because they can be geographically distributed and embedded in place through time. Taking a somewhat technocentric position, Goodchild (2007) also argues that for VGI to mature and extend further (with more people enrolled and performing more kinds of mapping) the barriers to entry must be lowered. This will be achieved, in part, by the routinisation of simple digital tools for data collection, embedded in the functionality of mobile devices such as smartphones, and online platforms for information authorship and collaborative knowledge creation becoming easier to use. He continues that VGI has value and should be nurtured because existing geographic knowledge about the world is incomplete and of variable accuracy.

3. Mapping for crisis preparedness and response

In order to understand the significance of mapping for crisis preparedness and response, we are briefly presenting the development of this concept in which new (open-source) ICTs and new actors are playing a crucial role. Crampton

(2009) stresses that it seems we are moving from a niche-based study of maps as objects to a more comprehensive (and potentially interdisciplinary) study of mapping as practice, the knowledges it deploys, and the political field of its operations. Neocleous (2003) states that maps are intended to convey the concepts or information intended by the dominant political forces, and a better question would ask how efficiently they do this. That some maps may be 'inaccurate' could well be part of their efficiency. Moreover Kitchin and Dodge (2007) argue that cartography is profitably conceived as a processual, rather than representational, science. They conclude that maps emerge through practices and have no secure ontological status. Maps are transitory and fleeting, being contingent, relational and context-dependent. Authors are convinced that such rethinking provides a fresh perspective on cartographic epistemology, and could work to provide a common framework for those who undertake mapping as applied knowledge (asking technical questions) and those that seek to critique such mapping as a form of power/knowledge (asking ideological questions). Crampton (quoted in Hall 2012) argues that mapping is a field of "knowledge and power relations" being pulled in several different directions. He describes a pull on one side toward "securitization" countered by "resistances" on the other. Securitization emerges from post-war efforts to liberate cartography of any association with art and propaganda and render it "post-political". However the resistance side of Crampton's diagram that we are interested in is characterized by critical cartography, map art and the open source movement meaning millions of amateurs and beginners using previously inaccessible mapping technologies to construct a vast "geoweb". A 2007 article in the New York Times on such practices optimistically described the amateurization of mapping as the "collective creation of a new kind of atlas" that would be "richer and messier than any other" (Hall 2012).

"Crisis mapping is a process of sourcing, visualizing and analysing data from a humanitarian, political, ecological, financial or other crisis, often in real time, on a dynamic, interactive map" (OCHA 2013, 61). Crisis maps are far from being new since they have already existed in the 17th century and later on used during first and second world wars to plan battles, to provide the information on military capabilities etc. However those maps were not updated in real time and they were controlled only by the few. In the contemporary era mapping is far more democratized since hundreds of live maps are being launched every week by ordinary people around the world. But this new type of geography is radically different from traditional approaches because it is user-generated and more participatory; the fact that today's dynamic maps can also be updated and shared in near real time opens up an entire new world of possibilities and responses, especially during times of crisis (Meier 2012, p. 90). As seen above crisis mapping originally defined as methods and practice to document armed conflict and human rights violations, is now more generally applied to armed conflict, disaster response, and complex emergencies. Moreover it should be noted that the methods and tools used in crisis mapping are often applied to non-humanitarian events as well (Payne et al. 2012).

Today's crisis mapping traces its origins to mapping and spatial analysis for humanitarian relief during crises. Relief agencies such as the United Nations Office for the Coordination of Humanitarian Affairs (UN OCHA) were searching for more efficient ICTs to assist in delivering emergency services. Agencies have long turned to GIS and remote sensing, for example, to map concentrations of victims, to inventory damaged roads, to calculate shortest paths to those affected by a natural or human-induced disaster, and to visually convey the crises on the ground (Kaiser et al., Poblet and Brandesescu in Brandusescu et al. 2015). The challenge in GIS was the steep learning curve in using these technologies, the reliance on top-down processes and experts, their consequent slowness in responding to emergencies as well as the disconnect of these remote experts to on-site activities. What gives crisis mapping its rhetorical power lies in its link of emergent easier-to-use mapping technologies with nonexperts (i.e. individuals not credentialed in humanitarian relief or in GIS). The earliest example of crisis mapping originated in Kenya in order to enable easy and fast reporting of postelection violence in 2008. Due to the dangerous situation on the ground that made local reporting difficult, the media coverage and direct information from local sources were inconsistent. Therefore the mapping–texting platform Ushahidi was established with the aim to address the need for the people on the ground to gain access to free mapping technology that allowed them to communicate with one another and with those seeking to help victims (Okolloh in Brandusescu et al. 2015). Ushahidi is accessible online and allows users to mark events on a map that changes in near real time. It has become one of the most commonly used and discussed tools for crisis mapping. It geo-locates reports sent to a website through SMS, the platform, together with its visualisation tool Crowdmap, translates, classifies and geo-references (using Google maps) reports sent, email and even voice messages (for the illiterate). Although Ushahidi was launched in Kenya, its best-known usage occurred during the 2010 Haiti earthquake. In the wake of the Haitian earthquake of 2010, Ushahidi set up a Haitian crisis mapping operation through which people and organisations posted their needs, and volunteers translated and picked up geo-located requests, and reportedly helped save many lives. Compared with the traditional, fixed map, two key distinctive characteristics of mapping like the Ushahidi project are that it is participatory and continually in flux (Hall 2012, pp. 158–159). Since 2010 several crisis in developing as well as developed countries were mapped such as refugee crise in Libya, Jordan and Somalia, earthquakes in Chile and Japan, floods in Pakistan, tornado in the USA state of Alabama and various others disasters. Despite the fact that crisis mapping is more frequently applied in developing countries, various crise in developed parts of the world have been mapped as well.

One of the examples were the explosions of three reactors at the Fukushima Daiichi power plant after 11 March 2011 disseminated nuclear radiation across the Japanese territory. Limited access to official information about the level and the location of radiation characterized the weeks following the event. However amateurs – nonexperts in geography and/or radiation-sensing – used online web-based mapping technologies, primarily Google Maps, attempting to locate the radiation and creating several online maps (Plantin 2015, pp. 904–905).

Crisis mapping is often carried out by already mentioned Volunteer and Technical Communities (V&TC), meaning individuals and groups outside the humanitarian community who use their expertise to collect, collate and disseminate useful information in a humanitarian response. V&TCs can range from untrained volunteers participating in crowdsourced exercises, to professional organizations or companies that devote resources to supporting humanitarian action or disaster response (OCHA 2013). V&TCs are offering services to humanitarian agencies for quite some years like MapAction offering mapping services, Télécoms Sans Frontières providing satellite-powered voice and data communications in emergencies, Sahana, an online humanitarian data-base providing 3W information about the stakeholders: who is doing what and where, etc. After the 2010 earthquake in Haiti, many new V&TCs were established and played a major role in the response such as OpenStreetMap (OSM), CrisisMappers, CrisisCamps/CrisisCommons, 4636 Alliance, Ushahidi (Disaster Relief 2.0. 2011). Meier (2012) has reviewed case studies in seven different countries with different types of humanitarian crisis that illustrate how the humanitarian space is moving from a traditional unipolar system to a more multipolar world order in which new actors are rising to the surface.

As highlighted by Meier (2012) such a radical shift from static, "dead" maps to live, dynamic maps, requires that we reconceptualise the way we think about maps and use them. Due to ICTs and (open-source) mapping technologies individuals/volunteers from all over the world are becoming an important actor in humanitarian response by being a source of a significant amount of real-time data via short message services (SMS; text messages) and social media platforms such as Facebook, Twitter, YouTube, Flickr, etc.

According to Andrea, Boccardo and Tonolo (quoted in Payne et al. 2012) we can separate volunteer geographic information into two categories: 1. Activities that collect and geolocate event reports and needs directly from the affected community via e-mail, short message service (SMS), and others, for example, use of the 4636 SMS shortcode and Ushahidi deployments; 2. Activities that develop primary geographic data over a disaster area. This includes, for example, the work conducted in the aftermath of the Haiti earthquake in January 2010 by volunteers mapping destroyed buildings through aerial photo interpretation. Since EUROSHA volunteers in our case study were conducting the second type of crisis mapping commonly done in close cooperation with a traditional humanitarian sector, we will briefly present these activities. They seek to leverage resources by moving data development and discovery from traditional information management to the crowd. These activities can be further classified as those that rely on any interested volunteer and those projects that require a certain level of geographic expertise in what is known as "bounded crowd-sourcing." Data arising from bounded crowd-sourced projects are more likely to adhere to international professional standards. The activities associated with OpenStreetMap (OSM) that has been applied in the framework of the EUROSHA as well, are particularly noteworthy when discussing crowd-sourced geographic data. OSM is an online live service of a global data set built entirely online by volunteers who may or may not have training in map development and is therefore unbounded. The Humanitarian OpenStreetMap Team (HOT) creates and provides free, up-to-date maps that can be a critical resource when relief organizations are responding to disasters or political crises. HOT applies the principles of open source and open data sharing for humanitarian response and economic development. The scale of the OSM project is the largest of its kind, with hundreds of thousands of volunteers worldwide creating data not only of streets but a large array of user-defined features including building footprints, parks, populated places, water bodies, pipelines and hundreds of other relevant geographic entities. OSM is often used as the base map for Ushahidi and other crisis deployments (HOT 2015; Payne et al. 2012). The OSM project is often considered the most successful collective map ever produced (Gerlach 2010; Johnson and Sieber 2012; Perkins 2007 in Caquard 2014), relying on dedicated contributors to compile a publicly accessible road map for the entire world at a very fine scale. It has been described as a way to develop new forms of communities around the world, based on shared values and practices (Lin quoted in Caquard 2014).

Among opportunities of creating maps for crisis preparedness and response, crisis mapping raises also different challenges. Literature is mostly emphasizing cartographic challenges such as how to provide a clear and unambiguous cartographic representation of a quickly changing situation on small-screen devices (Razikin et al. in Caquard 2014), how to help users to differentiate the origins and the quality of the data (Goodchild and Glennon; Roche et al. quoted in Caquard 2014), how to create information that is authored consistently, and with sufficient quality to match that produced by smaller numbers of paid professionals working in more conventional modes (Dodge and Kitchin 2012, p. 35). Moreover concerns are related to data validation (De Longueville et al. in Caquard 2014) and to the evenness of coverage, consistency and accuracy of content and metadata, along with the ongoing commitment and sustainability in labour exchanges without employment contracts and salary obligations (Dodge and Kitchin 2012, p. 35), etc. Study case of EUROSHA confirms such challenges together with others going beyond the cartographic ones.

4. European Open Source Humanitarian Aid (EUROSHA) Volunteers Project

The European Open Source Humanitarian Aid (EUROSHA) volunteers have been part of the pilot project titled "EU, Local and Online Volunteers: Key actors for inclusive humanitarian information-sharing in crisis preparedness", financed under the 2012 Call for Proposals for the European Voluntary Humanitarian Aid Corps (later renamed to EU Aid Volunteers). The project brought together nine organisations from five different European member states with expertise in humanitarian action, volunteering and Humanitarian Information Management (HIM) and 25 European and African volunteers. The combination of EU volunteering and open data for HIM, supported by a network of online volunteers, represents an ambitious dual innovation for the humanitarian and volunteering sectors alike. However the focus of the article will be on the project objective related to conducting "real" collaborative mapping for crisis prevention.

The main objective of EUROSHA was to develop a framework guaranteeing the added value and the security of European citizens' commitment to international humanitarian aid through volunteering in order to increase the effectiveness and inclusiveness of humanitarian aid. Therefore the project aimed at strengthening the humanitarian and volunteering communities' knowledge about the phenomenon of volunteer humanitarian information sharing, and about how it may improve the quality of humanitarian response. One of the specific objectives related to the research objectives of our study was to enhance the crisis-response environment by contributing to effective humanitarian information-collection and sharing solutions, thereby increasing the mutual visibility and capacity to collaborate between local, national and international crisis-management actors in deployment countries in preparation to crisis through open source tools and ICTs.

The project anticipated the selection, training, and deployment of European volunteers to countries where they have been working together with local volunteers, with remote technical support by HIM experts and online volunteers. Together, those volunteers have been facilitating crisis preparedness through geographic and humanitarian data collection, making data accessible through open-source websites (OpenStreetMap, Sahana Eden), information sharing, inter-actor linking, and capacity building of crisis response actors. Therefore the crowdsourcing, most particularly the mapping, as a relevant technique in the ICT era, was one of the most significant dimensions of the project. Since EUROSHA volunteers were training different relevant actors and handing over mapping activities, we must also stress the added value of crowdseeding. Crowdseeding is a related approach in which humanitarian organizations work with targeted people who are trained in gathering and sharing information. This allows for a deeper relationship between the aid agency and the data source. However in the particular case, the trainers were EUROSHA volunteers who have received comprehensive intense training, involving also GIS experts on mapping and particular open data platforms (OSM, Sahana). Further they have acted as "train the trainers" by providing trainings for Kenyan volunteers from different groups (students, local authorities, local and international humanitarian organizations, etc.) and tried to connect them with relevant humanitarian organizations. Therefore we can say that crowdsourcing in this particular case was "bounded". Besides trainees obtaining new knowledge and skills, and taking the responsibility for maintaining the mapping activities and upgrading maps, local volunteers have also offered great support in data collection and providing their accuracy due to the knowledge of local environment, language, people, etc. As stressed by OCHA (2013, p. 34) the accuracy of the information, meaning the probability that the information is a true representation of the situation at that moment in time is definitely critical in the ICT era where a vast volume as well as a complexity of information is available. Responders need to pay attention because if the basic data is wrong, decisions based on the analysis of that data will reflect those errors and can lead to incorrect results. Especially in cases of political unrests there is also a risk that different groups (civil-society groups, rebels, government, etc.) will establish different communication platforms delivering misinformation. Therefore while new technology increases individuals' ability to organize and promote change outside of formal structures, it also increases the capacity of governments or violent groups.

As stressed in the internal evaluation[1] of the EUROSHA project that was conducted by URD (Groupe Urgence, Rehabilitation, Developpement) in 2013, the project succeeded in responding to the emerging needs and challenges facing the humanitarian sector. The piloting of open source information management by EUROSHA volunteers responded to the crucial need for increased mutual knowledge and collaboration between local and international emergency management actors. The project demonstrated high potential and four teams of highly committed and professional volunteers succeeded in producing very satisfactory results that have gone beyond expectations on many levels. HIM activities have raised significant interest and prompted high levels of participation from a large number of organisations and individuals.

4.1 EUROSHA Volunteers in Kenya

The EUROSHA project has been carried out in four countries of deployment: Burundi, the Central African Republic, Chad and Kenya, as well as Cameroon later on. The choice of these host countries appeared relevant from the global as well as the European perspectives. Globally in terms of the relevance of crisis prevention, and linking relief, rehabilitation and development. Secondly since the countries of deployment feature complex and forgotten crises and have been selected as priority countries for the humanitarian strategy of Directorate General of the European Commission for Humanitarian Aid and Civil Protection (DG ECHO).

As also seen from the theoretical part of the article, of the four countries of deployment, Kenya probably had the highest technological potential, with higher than average computer skills amongst the local communities and wide availability of quite reliable 3G connection. It was also the only country with an existing OSM community (significant work carried out by Map Kibera Trust) and other strong mapping tools such as Ushahidi. However the EUROSHA volunteers were deployed to remote areas (Molo, Marigat, Njoro, Kuresoi) where these comparative advantages were somewhat compromised. However, they succeeded in convincing local partners of the usefulness and relevance of open data, not only in humanitarian contexts, but also for development activities (e.g. forestry management), which are very closely related to the prevention of future crises.

After a number of meetings and discussions, the EUROSHA Kenya team identified various topics related to possible development of crisis and thus important to be taken into account when planning the mapping process: floods, disputes due to the livestock, inter-tribal conflicts (Kikuyu, Kalenjin, Lou, Luhya, Kisii etc.),

1 The evaluation is internal therefore it cannot be accessible to the wider public.

ethnic clashes before, during and after the elections, displacements and establishment of settlements for IDPs. Moreover the team built close working relationships and partnerships with various actors such as local authorities (Dictrict Commissioners and District Officers for different sectors), international humanitarian and development organizations (World Vision, Red Cross, MSF Spain, Norwegian Refugee Council, OCHA, ICRC, Aga Khan Foundation), local NGOs (Network for Ecofarming in Africa (NECOFA), Map Kibera Trust (MKT), Maendeleo Endelevu Action Program (MEAP) and others. A special partnership was established with World Vision, which runs a large operation in Kenya. World Vision projects were mapped in the Marigat district and staff were trained in Nairobi and Marigat. World Vision decided to continue with the mapping activities and is prepared to mobilise other volunteers in the Marigat area who were trained by the EUROSHA team, when necessary. The EUROSHA Kenya team also worked very closely with NECOFA. In the first phase, volunteers mapped NECOFA projects and, secondly, staff were trained in mapping techniques. They are aware of the considerable advantages that mapping activities can provide when working with local communities and have subsequently written a project proposal which will include mapping. A special role was also played by Red Cross volunteers that participated in most of the trainings. All these partnerships contributed enormously to the success of the project. We can identify two main groups of achievements: enhanced capacities of various above-mentioned actors (around 50 individuals) that have undergone training on mapping and consequently the establishment of mapping community in rural areas of Kenya; the second are new or upgraded open-source maps. Figures 1 and 2 clearly present the extensive work that has been done by EUROSHA volunteers and their partners through crowdsourcing and ICTs.

Figure 1: Maps in Marigat before and after mapping

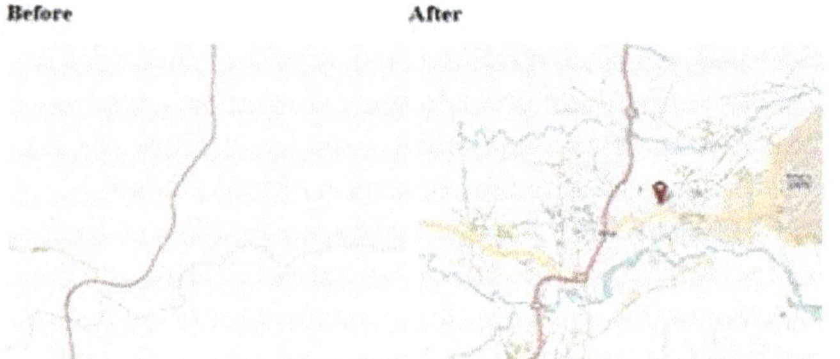

Figure 2: Maps in Molo before and after mapping

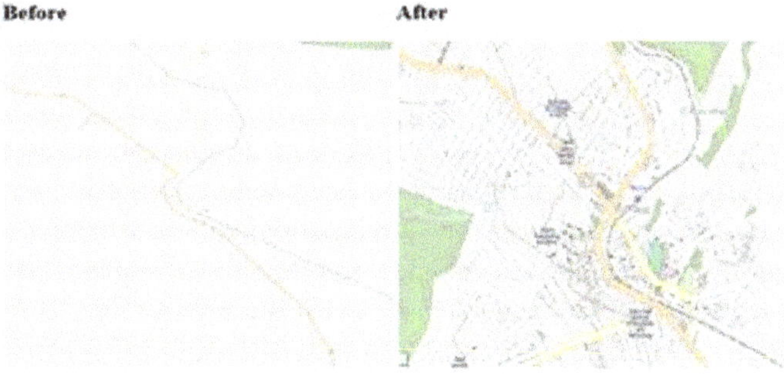

Besides opportunities and the above presented achievements, and the relevance of HIM/online volunteering there are also various constraints related to crisis mapping. Beginning with some of the challenges faced by the four groups of the EUROSHA Volunteers that are definitely relevant for numerous contexts of developing countries as well, we must stress that poor infrastructure in a couple of deployment countries, including frequent electricity cuts, limited computer access and internet connection, limited availability of other IT equipment and limited computer skills amongst local communities hampered progress and could have raised questions over the implementation and impacts of mapping activities. As such, it was likely that the project would produce more conclusive results in countries such as Burundi or Kenya, rather than in countries like Chad or CAR, and better sustainability. However, views diverge here given that good results were achieved in all four host countries and in Cameroon. Another aspect of this technological gap is that the volunteers did not feel sufficiently "prepared" for the fact that they were going to train people who had never used a computer before. Even though the volunteers were made aware of this challenge, more time could have been spent on this aspect during the training course and the volunteers could have benefitted from more tips on how to deal with this situation. Another issue that deserves reflection is the question of the use and sustainability of the software. We can conclude that crowdsourcing has definitely an important value, however the impact of this technique remains uncertain due to various reasons such as a need for strong partnerships and coordination, dedicated and properly trained and deployed volunteers, strong community or professional support, a rigorous approach to data validation, as well as an awareness of broader ethical and practical considerations, including privacy and security. As stressed by OCHA (2013) software serves to structure and streamline systems. When the structure is appropriate, software helps users and provides a lot of useful and easily understood services because users quickly understand the structure. If the structure does not perfectly match the user's expressed or unexpressed expectations, the user will feel frustrated and restricted by the software. The preliminary phase of software parameterization is fundamental to make it a positive experience. As a result, prior to parameterization and application, a thorough needs analysis is required. An in-depth understanding of the needs to be covered and the potential use of the system is essential to allow developers to build a structure which can provide a good service to the users. OpenStreetMap is a unique database which is available worldwide. However software which is not based on a single unique database like OSM can be installed and customised in many different locations for various types of user groups. As a result, each time a software instance is created for a specific usage, it is important to allocate resources to keep this instance running in the long term. These costs were not covered by the EUROSHA Volunteers project but they should be anticipated in the future. Moreover the volunteers in Kenya benefited significantly from the two visits made by the members of the HOT team that enabled additional training in the field, solved certain technical issues and supported them during the fieldwork. Also the remote support was very valuable from which we can conclude that bounded crowdsourcing has definitely a strong comparative advantage when it comes to the quality of mapping and maps. When it comes to Sahana, the EUROSHA volunteers produced certain good results, however many actors and stakeholders involved in the project, observed that the platform has not been used to its full potential, including in Kenya. Lack of training, time and dedicated resources meant that it was not possible to conduct a specific needs analysis for Sahana, nor provide proper continuous support for the volunteers during deployment. The user and role settings were developed without enough information on the context of usage. This resulted in a certain degree of frustration amongst the volunteers who found it difficult to carry out some tasks or who felt that the software contained bugs. These problems arose because primary parametrization was not done as it should have been and because it was not possible to provide enough support for the users who, once deployed, requested new features or modifications to the system.

Another important dimension that must be taken into account is the question of wide-scale information dissemination to the final beneficiaries (in the form of beneficiary-friendly products, including paper maps) as well as widespread two-way communication systems has been, and still is, a major unresolved issue in the field of HIM and GIS/Cartography. The two main questions is how can the process be improved to make sure that local communities receive the output (thematic maps, "normal" maps, etc.) in a form that they can actually use (online or printed maps) and how to encourage relevant actors to continue with mapping activities and contribute to the sustainability of the work and results. And last but not least when it comes to crowdsourcing in terms of volunteers collecting data in the field special attention must be given to the security and safety of the volunteers. To portray some of the moments indicating the importance of security, we can mention the attempt of burglary, suspicions from the police in Molo, mugging in Nairobi, resettlement of whole Kenyan team to Burundi as a pre-emptive measure in case of an escalation of insecurity in Kenya due to the elections. Contingency plans must be in place for such situations in order not to endanger the life of volunteers and hamper the implementation of the planned activities.

5. Conclusion

Taking into consideration the two parallel processes of growing and devastating natural and man-made disasters, and accelerating global connection of the world through ICTs, the ideas encompassed in the pilot project EURSOHA have a broader relevance. As concluded by the URD (2013) evaluation team the EUROSHA appears to be a significant project with very good results and high potential, both for the humanitarian sector and in terms of the volunteering experience that will be provided to a highly motivated group of participants under EU Aid Volunteers initiative 2015–2020. URD (2013) concluded that a volunteering framework bringing together actors from the volunteering sector, the humanitarian sector and HIM experts, could and should be replicated. We believe that these innovative activities are causing deep structural changes and have a great potential for improving HIM, especially due to the ICT and Web 2.0 era, and crowdsourcing participation of "citizen scientists." Moreover such crowdsourced mapping is significant since the state capacities and funding for increased costs of mapping are in decline under neoliberalism (see Dodge and Kitchin 2012; Leszczynski, 2012; Caquard 2014). Peck and Tickell (quoted in Leszczynski, 2012) talk about "roll-back"/"roll-out" neoliberalism, whereby the welfare state is either being eroded or actively rescinding itself from securing certain services and goods in the public sphere ("roll-back"), their provision is opened up to market regimes, which subsume the delivery of these goods and services to their own logics ("roll-out"). Leszczynski (2012, p. 73) suggest that "as "the state" is partially being rolled back, or actively withdrawing itself, from the public curation and dissemination of geographic information, markedly non-state (market and corporate) regimes of spatial data governance are rolling out in its place." However it must be emphasized that this does not mean the state is completely absent but that the relations between state, market and individuals in the field of mapping are being transformed.

Moreover as emphasized by various scholars (Dodge and Kitchin 2012; Caquard 2014, etc.) it is not necessary that the private sector will fill in these gaps, and what is particularly problematic is that the companies will not ensure uniform coverage of spaces since most probably they will focus on profitable markets such as urban street maps that are relatively easy to capture, package, and sell. Further the element of EUROSHA where volunteers have been sent to the field is also a valuable lesson as satellites are not able to detect many phenomena, especially not the details of phenomena, with remote sensing.

Besides opportunities of the ICT era for the humanitarian sector, there are also various constraints the responders must have in mind. ICTs definitely has the power to fill in the gaps of missing data, however the era has also brought increasing amounts of data that are hard to manage, especially in humanitarian situations with extremely limited time for understanding the situation and needs, and respond quickly. There can be participation biases related to the sources of information reflecting the different capacities of people and groups to produce information. The increasing access to ICTs and possibilities to inform the responders regarding the needs can also lead to increased expectations of people regarding the response that may go beyond realistic capacities. Without any hesitation we argue that the development of crowdsourcing crisis mapping is raising ontological as well as epistemological questions, especially related to the authorship that no longer relies on professional knowledge and experience, and the process and content that is constantly under review and is not being controlled by a single entity, especially not the state. Special attention must be given to the power relations. Even though ICTs enable the shift of power towards community, there can be many challenges related to influencing the local power dynamics. On one hand the traditional structures can be distorted and subverted, particularly in conflict-affected countries, by the provision of external resources, while on the other hand the local warlords can be entrenched when responders are being forced to act fast. Since we are talking about humanitarian settings, especially those overlapping with conflict settings, special attention must be given to the privacy of information or degree to which information should be public not to enable certain actors to use such information to undermine the security and safety of individuals, communities, humanitarian workers, etc.

When finishing the mapping mission in Kenya one of the open questions raising most doubts was the sustainability of work and results. We have concluded that attention must be definitely directed towards resources (appropriate equipment, internet access, etc.), knowledge and skills related to mapping and involving different relevant actors such as international and local humanitarian community, the civil society and technical groups, as well as establishing close cooperation with academic and government bodies. We believe that great support must be given to those who are showing most interest in continuing the mapping activities, and those who are deeply grounded in both local and civil society. They can provide a significant contribution, for example, by updating maps, providing online support, remote training or even the setting up of local associations of OSM. However we believe that in terms of sustainability the implementation of such ideas must be monitored and evaluated carefully in order to obtain relevant conclusion regarding the impacts of more inclusive humanitarian information management, new partnerships and mapping activities on the transformation of the humanitarian aid delivery.

The implementation of the EUROSHA project definitely confirms what was suggested by Dobson (2011), Meier (2011, 2012), Payne et al. (2012) that the most optimal approach in map compilation is a hybrid approach combining (1) traditional compilation techniques (e.g., field-work/field observation, data gathering/ data mining, use of imagery, conflation, data editing, data QC/QA) as practiced by staff with professional training in mapping/GIS, or by staff who have received training in map compilation techniques with (2) crowdsourced gathering of map data. The application of these new approaches taking advantages of ICTs are not intended to replace existing systems but enhance them and hopefully provide information that is reliable, time sensitive, and cost sensitive.

6. References

Brandusescu, Ana / Renée E Sieber / Sylvie Jochems: "Confronting the hype: The use of crisis mapping for community development." *Convergence: The International Journal of Research into New Media Technologies* 2013, pp. 1–17, retrieved 5.9.2015 from DOI 10.1177/1354856515584320.

Caquard, Sebastien: "Cartography II: Collective cartographies in the social media era." *Progress in Human Geography* 38 (1) 2014, pp. 141–150, retrieved 1.9.2015 from DOI 10.1177/0309132513514005.

Clifford, James: "Traveling cultures". In: Lawrence Grossberg / Cary Nelson / Paula Treichler (eds.): *Cultural Studies*. Routledge: New York and London, 1992, pp. 96–116.

Crampton, W. Jeremy: "Cartography: performative, participatory, political." *Progress in Human Geography* 33(6) 2009, pp. 840–848, retrieved 10.9.2015 from DOI 10.1177/0309132508105000.

Harvard Humanitarian Initiative / United Nations Foundation / UN Office for the Coordination of Humanitarian Affairs (OCHA) / Vodafone Foundation: *Disaster Relief 2.0. The future of information sharing in humanitarian emergencies,* retrieved 11.8.2015, from http://www.unfoundation.org/assets/pdf/disaster-relief-20-the.pdf.

Dobson, M.: *Google, Navteq and Map Compilation*, retrieved 10.8.2015, from http://blog.telemapics.com/?p=380.

Dodge Martin. / Kitchin, Rob: "Crowdsourced cartography: Mapping experience and knowledge." *Environmental and Planning* A 45(1) 2013, pp. 19–36, retrieved 2.9.2015 from DOI 10.1068/a44484.

Estellés-Arolas, E. / Gonzalez-Ladron-de-Guevara, F.: "Towards an integrated crowdsourcing definition." Journal of Information Science, 2012, pp. 1–14, retrieved 7.9.2015 from DOI 10.1177/016555150000000.

Goodchild Michael F.: "Citizens as sensors: the world of volunteered geography". *GeoJournal* 69 2007, pp. 211–221.

Hall, Peter: "Afterword: On Mapping and Maps." *Design Philosophy Papers* 10 (2) 2012, pp. 157–167, retrieved 23.8.2015 from DOI 10.2752/089279312X13968781797959.

Hörler, Raphael: *Crowdsourcing in The Humanitarian Network – An Analysis Of The Literature*. ETH Swiss Federal Institute of Technology Zurich: Zurich 2014, pp. 1–67.

Humanitarian OpenStreetMap Team (HOT), retrieved 1.8.2015, from https://hot osm.org/.

IRGC: *Risk Governance Deficits. An analysis and illustration of the most common deficits in risk governance*. International Risk Governance Council: Geneva 2009, pp. 1–92.

Kaiser R / Spiegel PB / Henderson AK, et al.: "The application of geographic information systems and global positioning systems in humanitarian emergencies: lessons learned, programme implications and future research." Disasters 27(2) 2003, pp. 127–140.

Kitchin, Rob / Martin Dodge: "Rethinking maps." *Progress in Human Geography* 31 2007, pp. 331–44, retrieved 8.9.2015 from DOI 10.1177/0309132507077082.

Leszczynski, Agnieszka: "Situating the Geoweb in political economy." *Progress in Human Geography* 36 (1) 2012, pp. 72–89, retrieved 5.9.2015 from DOI 10.1177/0309132511411231.

Meier, Patrick: *An open letter to the good people at Benetech*, retrieved 15.8.2015, from http://irevolution.net/2011/04/18/open-letter-benetech/.

Meier, Patrick: "Crisis Mapping in Action: How Open Source Software and Global Volunteer Networks Are Changing the World, One Map at a Time." *Journal of*

Map and Geography Libraries 8 (2) 2012, pp. 89–100, retrieved 11.9.2015 from DOI 10.1080/15420353.2012.663739.

Neocleous, Mark: "Off the Map: On Violence and Cartography." *European Journal of Social Theory* 6 (4) 2003, pp. 409–425, retrieved 5.9.2015 from DOI 10.1177/13684310030064003.

Nethope: *A case study series*, retrieved from 2.9.2015, from http://nethope.org/.

OCHA: *Humanitarism in the network age. Including world humanitarian data and trends 2012.* OCHA Policy and Studies Series. UN Office for the Coordination of Humanitarian Affairs (OCHA): New York 2013, pp. 1–113.

Payne, Karen / Patrick Florance / Stephen Shain: "The Role of Data Repositories in Humanitarian Information Management and Crisis Mapping." *Journal of Map and Geography Libraries* 8 (2) 2012, pp. 118–133, retrieved 14.9.2015 from DOI 10.1080/15420353.2012.662931.

Plantin, Jean-Christophe: "The politics of mapping platforms: participatory radiation mapping after the Fukushima Daiichi disaster." *Media, Culture and Society* 37 (6), 2015, pp. 904–921, retrieved 1.10.2015 from DOI 10.1177/0163443715584102.

Stephenson, Max / Marcy Schnitzer: "Exploring the Challenges and Prospects for Polycentricity in International Humanitarian Relief." *American Behavioral Scientist* 52 (6) 2009, pp. 919–932, retrieved 18.9.2015 from DOI 10.1177/0002764208327667.

Stewart, Alex: The Ethnographer's Method. Qualitative Research Methods Series 46. Sage: Newbury Park 1998, pp. 1–97.

URD: *Internal evaluation report. EUROSHA volunteers project: EU, local and online volunteers: Key actors for inclusive humanitarian information sharing in crisis preparedness.* Groupe URD: Plaisians 2013, pp. 1–61.

Changing Culture Dynamics: Identities, Knowledge and the Power of Networks

Tea Golob
School of Advanced Social Studies in Nova Gorica
Gregorčičeva 19, 5000 Nova Gorica, Slovenia
tea.golob@fuds.si

Instrumental Identities and Information Society – Deploying the Issue of Social Fields

Abstract: *The chapter addresses the information society as a concept referring to recent transformations of the social reality expressed in the form of semantic structures. It emphasizes that an excessive number of information does not per se indicates that all people have the same access to that information or that they are equally able to contextualise them. Different access to resources, which can be cognitive or psychical, is playing a crucial role in that regard, which has a significant impact also on contemporary identity constructions. Individuals can form either strategic instrumental identifications, which serve as a mean of successful adaptation to social settings and also of transforming them, or they can passively absorb the transformed cultural information, which influences their identity construction without strategic endeavours. In order to represent a meaning of different national setting in substantiating contextualisation of increased information and accordingly strategic identifications, the text draws on a fuzzy set method as designed by Charles C. Ragin (2000). The analysis elucidates that while social environment has become more differentiated, it has started to offer new semantics on a systemic level, which increasingly trigger individuals' cognitive maps. Individuals have become more reflexive and less bounded with social structural determinants. They are able to adjust their personal semantics referring to self-description and belongings more intensively. Accordingly, they can form multi-layered, instrumental identifications, not necessarily tied with their primary role-sets but those identities consequent different life-paths and goals, which someone tends to achieve.*

Keywords: *information society, reflexivity, social fields, identities*

1. Introduction

As the title of the chapter implies, at stake are identities, which assumingly take over specific meaning provided by the new dynamics of the social order, which has been described for instance as a network society (Castells 2000), liquid modernity (Bauman 2000), an information revolution (Morozov 2013), etc. One can notice an excessive bricolage of different theoretical approaches (cf. Webster 2014) dealing with various dimensions stemming from the increased amount of information and technologies enabling a dissemination of this information. Herein, the attempt is not to emphasise the new epoch of the information age as being drastically differentiated from other social orders. It has been even argued (cf. Webster 2014) that the information society does not indicate social settings *sui generis,* although different ways of social organisations, communications and individual perceptions have emerged. While recognising certain unignorable transformations of the social realities referring to technological development and mass mediation, the central goal is to represent that certain characteristics of presumingly declining social imperatives are still present.

Considering the information society as a concept referring to recent transformations of the social reality, the usefulness of the concept lies in recognition of the role that information has on the semantic structures. Due to the global spread of massive information, social order has become more complex, differentiated and less stable. The role of reflexivity on the systemic and individual level has thus been reinforced. However, the excessive number of information does not per se indicate that all people have the same access to that information or that are equally able to contextualise them. As Webster says, there is "an important distinction between having information and being informed" (2014, p. 32). The expanding complexity of information reflecting the increasing complexity of social structure can be reduced by deliberate considerations of individuals, which can thus strategically adjust social context to better suit their needs and concerns. We believe that it is not pretentious to claim that only the minority possess the ability of strategic actions and of regularly steering their life-paths. Different access to resources, which can be cognitive or psychical, is playing a crucial role in that regard. The latter becomes quite evident in exploring the issue of contemporary identification processes.

The massive flow of information erodes traditional role-sets in identity constructions to a certain extent, as exposure to cultural influences from disparate parts of the world unavoidably cause the transformation of rooted images of the social reality. Or as Appadurai (1996) says, due to globalisation and accompanied processes, people have started to deploy imagination in every aspect of their lives, which has thus become a collective, social act. Accordingly, the perceptions of our-selves entail those imaginations. However, he also recognises that imaginations or aspirations are not evenly distributed around the world, and they are formed in "[the thick of social life]" (Appadurai 2004, p. 67). Identities have been changing, and this is indisputable. However, individuals can form either strategic instrumental identifications, which serve as a mean of successful adaptation to social settings and also of transforming them, or they can passively absorb the transformed cultural information, which influences their identity construction without strategic endeavours.

Individual's meaning making of the self and the others, which leads to identifications, is far from straightforward. On the one side, there are structural forces and on the other are individuals, with their inner spaces and intimate revaluations. There is interplay between both levels, which has attracted the attention of scholars from the very beginning of scholarly thought. In order to understand that interplay, one needs to define the relation between both levels in the first place. Herein, we follow perspectives of critical realist and especially of Margaret Archer (1988; 2003; 2012). The most important premises of such thinking, which were systematically recognised by Mingers et al. (2013, pp. 794–795), are that there is a causally efficacious world, which exists independently of our constructions and knowledge. Critical realism advocates that there is neither the reality, which can be reduced only to empirical observations and measures as implied by classical positivism, while it also defends against constructivism conditioning the existence of the world with human knowledge. It also recognises that our relation to the world is always resulting from evaluations ensuing from our perceptual and theoretical lenses. It gives credence to epistemic relativity that knowledge should be considered through the lens of its local and historical dimensions, but it advocates against judgmental relativity referring to existence of equally valid views and opinions. Furthermore, critical realism also emphasises that there are three categories of objects referring to knowledge, therefore entailing physical, social, and conceptual sets. Recognising the latter requires a deployment of various methodological techniques and concrete methods. While advocating a dialectical relationship between individuals and social structure, in the fore are not ontological differences between emergent properties of both entities as such, but the need for methodological division. The latter is required in order to recognise the co-influence between both, which lead to analytical dualism (Archer 1988).

Therefore, we consider human persons (psychical systems) and society (social systems) as two emergent entities, which cannot directly influence each other. Social structure and individuals present emergent entities, which should not be equated. Social structures, entities or systems cannot exist without people; however, they could not interfere with each other's projects directly. The mutual influence between both levels is thus inter-reaction between causal powers of both emergent properties (i.e. of social and psychical systems). Identities reflect the belongings and self-recognition of individuals within particular social contexts, but the influence of the latter is a result of the complex interplay between individual deliberations and social forces. Following Archer (2003; 2012), we bring to the forefront the ability of reflexivity, which mediates the social context and individual concerns. This reflexivity is an inevitable part of human existence but, due to global social transformations, one can see that its role has been reinforced. Individuals cannot consciously put into process all the information that social context comprises, but are nevertheless forced to increase the intensity of them.

Each person has her or his own inner place in which she or he is able to reflexively deliberate upon the social context and thus construct identities, but she or he does not interfere into social context directly. Individuals' agential powers evoke particular social emergent properties (i.e. enablements and constraint) on which they can deliberate. Saying that the exercise of enablements and constraint is contingent upon personal agency (Archer 2003), one needs to orient her or his concerns to particular social contexts in the first place. And there is a crucial role of different resources conditioning our social embeddedness. The lack of resources has an impact on someone's concerns as a result of reflexive deliberation. A variety of resources (i.e. material, social and cultural) conditions the individual's embeddedness in particular social context, which also expresses involuntary agential positioning. Social embeddedness determines social imperatives defined by social structure, which provides the reproduction of social systems. Social imperatives are seen as potential causal powers of structural emergent properties, which have a tendency to impinge upon individuals, but can also remain unexercised. They are seen as enablements and constraints, which imply individual agency (Archer 2003). Social imperatives refer to discursive influences, but also to non-discursive ones. Following Archer (2003) and the perspectives of realists, the meaning produced by social communication rests on a Peircean foundation. In that regard, the important role is played by the recognition of the triadic model of the sign divided on *representamen, interpretant*, and *object* to which the sign refers. Besides, as Elder Vass says, there is also its "[relation to extradiscursive referents and an extracommunicative world' that should be taken into account]" (2007, p. 428). As we are interested in the determination of individual's social embeddedness on the discursive influences acting as a potential trigger on his or her reflexivity, we define the scope of discursive influences as codes. The inspiration can be found in Bernshein's restrictive and elaborated codes, but only as an idea that there are different ways of saying, with consequences for different ways of meaning. Therefore, codes are a representative of semantic variation (cf. Hasan 2009). The latter can be a potential trigger to an individual's cognition but due to the particular aspects of social embeddedness, individuals can also be a medium of dissemination. It also needs to be said that Archer recognises the role of non-conscious features in the depiction of the passive agents, i.e. individuals to whom things happen (Archer 2003). In the individual's reflexivity, an important role is played also by "interactional" and "figurational" inter-relations between individuals (Mouzelis 2007), of which constellation is defined by the social environment as the "institutional game", and are crucial especially in intersubjective meaning making (Bottero 2010). The listed social imperatives impinge upon psychical systems and their internal conversation, which can lead to reflexive deliberations, of which transnational identifications are part.

In transnational social spheres, social imperatives more often act as a trigger to an individual's cognition. Reflexivity increases or can be related to a communicative, autonomous or meta-reflexive modes of reflexivity using Archer's formulation. Particularly, while the last two modes are exercised, the identifications as a project enable morphogenesis, which is progressive and reflects the actors' ability to influence the social settings according to their internal conversation. For individuals who stay within national frames, it is more common that they are predominantly a medium of dissemination. They can remain in their involuntary agential position and mediate social and cultural emergent properties without changing their positions in social or cultural structures. They are involved in the primary agency or the mode of reflexivity can be communicative. The identifications, even if they are a product of reflexive deliberations, contribute to morphostasis. In the contemporary era of late modernity, the latter situation should be rare and less deterministic. However, not all can regularly participate in the transnational social sphere and be able to contextualise the flood of information in the same way.

The social spheres, be they national and transnational, are perceived as social environments in a broader sense, not limited to physical spaces or communities. For instance, one who is participating in a national social sphere is not necessarily living within borders of a particular national country. An important role is played by his or her daily practices expressing or communicating a prevailing form of socially constructed collectivity. As Appadurai suggests, grasping the new role of imagination, we need to bring together the idea of the imagined community (e.g. Anderson 1983) and the ideas of the imaginary as a constructed landscape of collective aspirations, which are becoming intensively involved in mobility and connected with transnational social and cultural processes (Apadurai 1996, p. 31). He offered the perspectives of global scapes, which are not "[objectively given relations that look the same from the very angle of vision]", but imagined worlds of historically situated imaginations of persons and groups around the world (1996, p. 33). Their embeddedness in different political, cultural, economic and social contexts conditions their encounter of flows, from which they construct the worldview (Appadurai 1996, p. 33). The differentiations of national and transnational spheres refer to the context and the flows influencing the imagination of someone's social environment. Although the world is inevitably interconnected in all aspects of social life, the latter is not necessarily equally present in imagined worlds. Each type of a sphere (national or transnational) expresses a predominant variety of social forces referring to discursive codes and figurational relations or, as Beckert (2010) says, institutional rules, network topographies and cognitive frames. A national social sphere signifies structural properties predominantly limited to a national level, while a transnational or supranational sphere reaches beyond national boundaries. All three social forces are a part of the interplay between potential structural properties and reflexive deliberations.

2. The Concept of Social Fields and Theoretical Framework

In order to approach the process of contextualising and absorbing information in the recent era, the concept of (transnational) social field can be of great help. The concept is often associated with Bourdieu (1990), who saw fields as social settings, in which individuals draw up an objective social structure into their personal dispositions and they act in accordance to the absorbed social context. The latter is revaluated through habitus reflecting the different types of capital. However, the role of individual reflexivity and intimate deliberations seems to be neglected in that regard. In order to reach beyond overly deterministic perspectives, the integrated theory of strategic social fields proposed by Fligstein and McAdam (2012) can advance our discussion. They perceive social fields as local orders, where interaction between incumbents and challengers according to shared rules and meanings take place. Dynamics within a particular field can be seen as an intermediary level between structures and agents, which make social fields to be a useful concept for exploring identifications in a social world where social, economic and cultural connectivity has increased, and the nature of individual consciousness, people's motivation to pursue their ends, and the extent of their freedom from external influences have become a challenging theoretical and empirical task. According to Fligstein and McAdam (2012), there is infinity of strategic social fields, of which some can be institutionalised, or nested within other fields. The national social fields can be seen for instance as a framework of many other strategic fields, which can also be interacting beyond boundaries of a nation state. According to transformations of the social settings induced by transnational connectivity and the massive spread of information, the issue of social fields has also been transferred to international and transnational spheres (initiating with Fouron and Glick Schiller 2001; Levitt and Glick Schiller 2004). Conceptualisations of transnational social fields do not differ much from that proposed by Fligstein and McAdam who see fields as local social orders or social arenas, because as Levitt and Glick-Schiller argue, "[in one sense, all (social fields) are local in that near and distant connections penetrate the daily lives of individuals lived within a locale]" (2004, p. 10). However, the concept embraces the simultaneous participation of individuals in local, national and also transnational social spheres. Social forces within one social field, seen as institutional rules, network topography and cognitive fames (cf. Beckert), are on a transnational level endowed with transnational dimensions expressed also in different types of capital, which are transnational in their nature.

We have argued elsewhere that active participation in transnational social fields enables individuals to form instrumental transnational identifications (cf. Golob 2014, Golob et. al. 2014). Those identifications express the reflexive deliberations of the social context and also the strategic use of different types of information leading to actions and to adjustments of social settings. The ability to actively participate in a transnational social sphere, contextualising and not just receiving information and thus inducing strategic actions, depends on different conditions representing an array of individual and collective backgrounds. What seems to be important, is that individuals need to possess a certain amount of resources (or capitals) while being embedded in the national social framework in order to enter the transnational social spaces and to become a part of transnational strategic action fields. For instance, in the case of European social fields, which are transnational in its nature, there are more educated, white-collar, predominantly young and wealthier people, who participate there (cf. Fligstein 2008). Or, as it has been shown in a case of Swiss international business elites, these actors are still part of specific national elite fields that are characterized by a series of structural relations between positions that are differently equipped with capital (cf. Bühlmann et. al. 2013). It is economic, cultural and social capital that seems to be crucial, while the amount of a particular type of capital can vary. In the case of managerial elites, those individuals need to be well educated, have extended social networks, and possess significant economic resources in order to participate in the transnational social sphere. In contrast, in a case of a transnational diaspora, the economic capital may not be so important, while (transnational) social networks and (transnational) cultural capital in terms of narratives of common origin are crucial. Individuals incorporate their own histories referring to a particular local and national environment into the transnational sphere, where they interact with each other. They come into contact with a variety of social memories, narratives and values that induces a contested set of (transnational) cultural information encouraging reflexive deliberation and consequently a formation of instrumental identifications. As mentioned, social fields are also nested in a complex web of other fields, and there is a distinction between distant or proximate fields, dependent and interdependent field or state and non-state fields, which is important (McAdam and Fligstein 2012). Participation in transnational social fields is marked by the individual embeddednes in other (also national) social fields, which substantiate certain lifestyles, individual concerns and actions. An important issue here, therefore, also refers to different backgrounds of individuals, which reflect the influence of all three mentioned social forces and the role of different types of capital, which is always field-specific (Bourdieu 1990).

In the present chapter, we would like to focus on the factors substantiating the platform encouraging or enabling the participation in the transnational social sphere. Although the issue of identification is to be explored on the level of individuals, herein, we would like to add a puzzle to the otherwise more comprehensible story. Thus, we intend to elucidate certain aspects referring to a macro level and to national contexts conditioning transnational instrumental identifications. The focus is on factors influencing the general framework of national social fields and the amounts of capital, which members of national community have on their disposal. We believe that not just intimate characteristic and personal backgrounds of individuals influence the ability to enter the transnational social sphere, but also national structural constellations. Individual properties, which are expressed in a certain mode of reflexivity (Archer 2012) or understood as social skills denoting what people do to attain collective action and give meaning and sense to our lives (cf. Fligstein and McAdam 2012, p. 46), depend on personal possession of particular resources. On the other hand, the national social structure indicates a general aggregation of resources potentially influencing personal ones.

In order to represent a meaning of different national setting in substantiating contextualisation of increased information and accordingly strategic identifications, we draw on a fuzzy set method as designed by Charles C. Ragin (2000). The method allows the macro-level research on a small number on units, and also provides complex causal relationships between different factors. It can reveal necessary and sufficient factors entailing the complex relationality. The factors considered to be important in substantiating the framework of national social fields are: general level of development of a nation state; integration into the global flows in terms of a level of globalisation of a country, national aggregate of individuals participating in a transnational social sphere; aggregate of a level of informatization of households on a national level; national aggregate of cosmopolitan values; and post-material values. Those are considered to be independent variables.

Those variables also implies different types of social resources or capitals, which define national social settings and also indicate the ability of entering the transnational social sphere composed of the transnational social fields. Those social resources are inextricably connected to asserted social forces although the boundaries of each category do not necessary overlap. For instance, social capital is linked to network topography, cultural capital refers to cognitive frames, while institutional rules embraces economic, political and also other types of capital. Therefore, transnational social practices refer to social capital and while being an aggregate of individual actions thus indirectly refer to network topography. Values refer to cultural capital and indirectly to cognitive frames, while the level of development is, for instance, simultaneously an expression of cognitive frames and institutional rules as, by being measured by Human development index, it entails the educational level and also economic resources of a country. Also the level of households' informatization is a complex intertwinement of institutional framework, cognitive frames and network topography and refers to a variety of capitals, which one can have on disposal. As a dependent variable expressing the most proper indicator of instrumental transnational identifications, we propose to be a strong feeling of European citizenship.

In order to operationalize the factors of influence in a particular national social environment we assert the following model.

3. Testing the Model

The study focuses on seven countries of the European Union, therefore Bulgaria, Denmark, Slovenia, Hungary, Sweden, Portugal, and Netherlands that are part of certain regions referring to North-South division and also to West-East distinction. As we used different databases, of which not all include equal countries, we decided to choose countries, which are a part of different parts of Europe and also assessed the membership of the EU in different time periods. Therefore, Sweden and Denmark are a part of Northern Europe, the Netherlands is a part of the West, Portugal is a part of South Europe, Slovenia and Hungary represents members of the former communist bloc and they are both placed in middle-east Europe, while the Bulgaria is a part of the east Europe.

Figure 1: The Model

<table>
<tr><th colspan="3">National social field as a framework of nested strategic social fields</th></tr>
<tr><th>Institutional rules</th><th>Cognitive frames</th><th>Network topography</th></tr>
<tr><td colspan="3">Social forces assert the general amount of different resources provided by national social environment, which influence individuals' ability of entering transnational sphere → contextualising flows of information → constructing instrumental identifications</td></tr>
<tr><td colspan="3">Resources in terms of economic, cultural and social capital</td></tr>
<tr><td colspan="3">Operationalization of resources – independent variables</td></tr>
<tr><td>Economic capital
Human development index 2013, (dev)
Index of globalisation KOF 2011 (glo)</td><td>Cultural capital
Human development index 2013; (dev)
Post-material values EVS 2008, (pm)
Cosmopolitan values EVS 2008 (co)</td><td>Social capital
Informatization of households, special EB 396 2013 icc;
Transnational social practices, EB 82 2014 tp;
Index of globalisation KOF 2011 (go)</td></tr>
<tr><td colspan="3">Instrumental identifications (euid) – Operationalization – measured by – definitely feeling as European citizen EB 81 2014</td></tr>
</table>

4. Explaining the variables

4.1 Development

Level of development entails a complex variety of social dispositions enabling individuals to enter the transnational sphere and to construct instrumental identities. Measured by Human Development Index (HDI), the variable is composed of different indicators, which expresses the economic growth, social prosperity and common well-being. A high level of development denotes more spare time enabling individuals to travel and to socialise beyond national boundaries. It also reflects the level of education played in a certain national environment. The increased role of education actually expresses the transition from traditional, industrial societies to post-industrial ones (Bell 1973). Accordingly, development determines the changes in value systems. The information society is heavily dependent on individual motivations and efforts, which are, together with other forms of education, accompanied with knowledge of foreign languages and cultures. Education has been even recognised as a potential for "transnational competence" (Mau and Mewes 2012).

4.2 Transnational practices

Regular transnational practices express the participation in transnational social networks. They entail the attachment to a transnational social sphere, which induces cultural exchanges, solidarity among people and can thus also forge feelings of belonging referring to a transnational social scale. Exposure to transnational social encounters and cultural influences can erode the traditional rooted cultural information and stimulate reflexive deliberations of individuals' concerns. In order to regularly participate on a transnational social level, individuals are obliged contextualising the flood of information offered by social settings as the presence on that scale demands a transition from national frames, which demands an active person. The passive persons, therefore ones to whom things happen, are not able to orient their actions reaching beyond local or national levels. In terms of measurement, transnational participation is referred from particular attitudes towards transnational connectivity, from transnational social networking and actual transnational participation. The parameter consists of three indicators, namely the percentage of people who visited an EU country in the last 12 months, those who read a book in another language or socialised with people from another EU country.

4.3 Globalisation

The level of global integration of particular national society denotes a complex variety of impacts that this integration can have on a national environment. Herein, we measure the indicator by drawing on the KOF: Konjunkturforschungsstelle globalisation index, which comprises three dimensions of globalisation, therefore, economic, social and political.Each dimension expresses a specific set of influences caused by global connectivity. On the economic level, it can be linked to development of a particular country, as it entails the data on the actual flows of the trade summing exports and imports of goods and services measured as a share of gross domestic product, further it takes into account foreign direct investments, portfolio investment and income payments to foreign national (Dreher 2006). Economic globalisation thus has an influence on a general amount of economic capital offered by the national environment, to which members of that environment can have a potential assess. The social dimension of the indicator refers to personal contacts on the global level and also on the information flows embracing internet user and media dissemination. The indicator also takes into account cultural parameters such as data on cultural proximity (Dreher 2006). Thus it signifies the potential amount of transnational social and cultural capital in a particular national environment or the conditions for its emergence. The political dimension of the indicator embracing international treaties, participation in international communities, missions and organisations (ibid.) reveals the social force of institutional rules, which constitute the structure of the national social field. More globally or transnationally oriented institutional rules might have a potential impact on the individuals' actions within a field.

4.4 Cosmopolitan values

Cosmopolitan values are an indicator of supranational values, opposing traditional national conceptions of exclusive belongings and dichotomous revaluation between 'us' and 'them'. It has been argued that identities of those, who are recognised as cosmopolitans, are open, reflexive and more inclusive (Vertovec and Cohen 2002). The presence of those values implicates the existence of cosmopolitan capital (cf. Bühlmann 2013), tightly knitted with other transnational resources. The indicator thus reveals the state of openness of the national social environment towards the social spheres reaching beyond national boundaries. The indicator herein is composed of three parameters, therefore the proportion of people that reject immigrants, Muslims or people of other race as being their neighbours.

4.5 Post-material values

Processes of identity construction and perceptions of ourselves in relation to others reveal a wider constellation of value and moral standards of particular society. Post-material values are intertwined with transformed perspectives on social reality induced from increased complexity, individualisation and post-modern relativisation of meanings. Post-modern values emphasise the importance of self-expression and simultaneously supporting tolerance and acceptance of other social groups. As Delanty says (2002), postmodernity calls into question the universality of identities and thus enables the discourse of multi-layered and multiple identities, which reach beyond the national environment. Post-material values express certain changes in values systems and orientations, which encourage participation in a transnational social sphere and allow transnational instrumental identifications to emerge. The post-material values are induced from the index of post-material values calculated from prevailing values in a particular national society.

4.6 Supranational transnational identifications

Identifications as a process, and their impact for individuals' life-paths, are most appropriate to be understood through qualitative in-depth studies. Finding a proper indicator on a macro level is thus always a hard task to be done. Therefore, our research draws on previous studies, which attempted to revaluate the indicator of instrumental identification on a macro level (details in Golob et. al. 2014). Thus European citizenship is not perceived as merely a status, but as indicators referring to thin cultural information, which is rational, endogenous and individual (cf. Mishler and Pollack 2003). As a proper measure we take an indicator of European citizenship, while a distinction is made between those who claim they definitely feel that they are European citizens and the rest (who claim they only feel this to some extent, not really, not at all, or do not know). Herein, only those who definitely feel as an EU citizen are taken into account.

Table 1: The values of the parameters

states	euid	glo	devel	pm	co	tp	icc
BG	14	71,38	0,777	-40	19,2	8,33333	44,33333
DK	34	87,43	0,9	7	8,1	48,3333	70
HU	25	85,91	0,818	-25,7	11,93333	8,33333	48,3333
NL	23	91,33	0,915	8,1	13,83333	50,33333	72,3333
PT	15	87,01	0,822	-34,3	10,7	11,66667	41,3333
SI	19	76,86	0,874	-7,2	28,9	35	58,66667
SE	32	87,39	0,898	11,9	9,7	47	71,66667

5. Testing the impact of proposed variables

The fuzzy set method allows for proposing data, which are not limited to dichotomous 'crisps'. Raw data on the selected variables should thus be calibrated into scores of membership in the qualitatively defined conceptual sets. The calibration underwent two phases. In the first phase, the fuzzy scale was selected. It consists of three qualitative anchors. The three-value scheme uses four numerical values 0 (with verbal label fully out of the set), 0.5 (with verbal label of half-in, half-out), and 1.0 (with verbal label fully in the set).

By using the computer programme fs/QCA 2.0, we were able to conduct the analysis of necessary and sufficient conditions contributing towards the presence or absence of a certain outcome. In the present situation, we were predominantly focused on the calculation of the necessary and sufficient conditions for the intensity of instrumental identifications in a particular national environment. Ragin (2000) defines the necessary conditions of specific social phenomena as those conditions whose absence automatically means that particular social phenomena will not occur regardless the presence of other conditions. A necessary condition is by itself capable of leading towards a researched result/outcome (Ragin 2000, p. 91). Sufficient conditions are those conditions that need the presence of at least one other condition to trigger the result/outcome. Sufficient conditions offer a challenge to a researcher since they highlight the possibility of the existence of at least one undiscovered condition.

5.1 The anchors and the calibration phase

compute: euif = calibrate(eui,34,24,14)
compute: glof = calibrate(glo,91.33,81.35,71.38)
compute: develf = calibrate(devel,0.915,0.821,0.777)
compute: pmf = calibrate(pm,11.9,-14,-40)
compute: cof = calibrate(co,8.1,18.5,28.9)
compute: tpf = calibrate(tp,50.3333,29,8.3333)
compute: iccf = calibrate(icc,72.33333,56.5,41.3333)

Table 2: Final fuzzy values

states	euif	glof	develf	pmf	cof	tpf	iccf
BG	0,05	0,05	0,05	0,05	0,45	0,05	0,08
DK	0,95	0,86	0,93	0,92	0,95	0,94	0,93
HU	0,57	0,8	0,45	0,21	0,87	0,05	0,17
NL	0,43	0,95	0,95	0,93	0,79	0,95	0,95
PT	0,06	0,85	0,51	0,09	0,9	0,07	0,05
SI	0,18	0,21	0,84	0,69	0,05	0,7	0,6
SE	0,92	0,86	0,92	0,95	0,93	0,93	0,95

5.2 Analysis of sufficient conditions

****

TRUTH TABLE ANALYSIS

****

File: D:/fuzzy values.csv

Model: euif = f(glof, develf, pmf, cof, tpf, iccf)

Rows: 5

Algorithm: Quine-McCluskey

True: 1

--- COMPLEX SOLUTION ---
frequency cutoff: 1.000000

consistency cutoff: 0.867159

	raw coverage	unique coverage	consistency
	----------	----------	----------
glof*develf*pmf*cof*tpf*iccf	0.743671	0.743671	0.867159

solution coverage: 0.743671
solution consistency: 0.867159

Based on the consistency criteria (exceeding the 85% threshold), all the factors combined together can be seen as sufficient condition for the presence of instrumental identifications. The coverage is lower but the reason can be found in the complexity of the statement combining six indicators together. Further, we conducted an analysis of necessary conditions, which revels the following:

Outcome variable: euif

Table 3: Conditions tested:

	Consistency	Coverage
glof	0.952532	0.657205
develf	0.955696	0.649462
pmf	0.876582	0.721354
cof	0.958861	0.613360
tpf	0.832278	0.712737
iccf	0.863924	0.731903

All the indicators, except the transnational practices, are also necessary conditions, which emphasise the relevance of the defined parameters. The level of globalisation, development and also cosmopolitan values has very high consistency, while also the transnational practices almost exceed the threshold.

6. Conclusion

The information society demarcates a complex concept, which can be understood in various ways. Herein, the focus is on the influence that information has on the semantic structure of the social order and also on individuals, being an auto-poetic psychical system having their own emergent properties. While the social environment has become more differentiated, it has started to offer new semantics on a systemic level, which increasingly trigger individuals' cognitive maps. Individuals have become more reflexive and less bounded with social structural determinants. They are able to adjust their personal semantics referring to self-description and belongings more intensively. Accordingly, they can form multi-layered, instrumental identifications, not necessarily tied with their primary role-sets but those identities consequent different life-paths and goals, which someone tends to achieve. The process of identity constructions thus results from personal deliberations and intimate revaluations. Nevertheless, their social embeddedness plays a certain role in that regard. The social environment offers a social playground, where rules and consenting values assert the rules of the games, which can be redefined in order to meet someone's expectations. Redefining the rule, however, depends on the amount of resources or capitals that one possesses and also has at his or her disposal. The national environment can have an enabling or constraining effect on someone's deliberate decisions. It does not present the imperatives of social actions in total but offers the certain limitations or encouraging impulses toward more reflexive concerns of individuals. It is not an attempt to delimit individuals' actions and their personal creativity to the imperatives caused by social settings, but to show the complexity of the interplay between systemic and individual level. The information society as a contemporary phenomenon and as a concept offers a productive venture point in advancing the on-going debate on social and individual emergent structural properties and their structural coupling.

7. References

Appadurai, Arjun: *Modernity at Large: Cultural Dimensions of Globalization*. Minneapolis: University of Minnesota Press. 1996.

Appadurai, Arjun: "The Capacity to Aspire: Culture and the Terms of Recognition". In: Vijayendra Rao / Michael Walton (eds.): *Culture and Public Action*. Stanford University Press. 2004, pp. 59–84.

Archer, Margaret: *Culture and agency*. Cambridge, New York, Cambridge University Press. 1988.

Archer, Margaret: *Structure, Agency, and the Internal Conversation*. Cambridge, New York, Cambridge University Press. 2003.

Archer, Margaret: *The Reflexive Imperative in Late Modernity*. Cambridge: Cambridge University Press. 2012.

Bauman, Zygmunt: *Liquid Modernity*. Cambridge: Polity. 2000.

Beckert, Jens: "How Do Fields Change? The Interrelations of Institutions, Networks, and Cognition in the Dynamics of Markets". *Organization Studies* 31(5), 2010, pp. 605–627.

Bell, Daniel: *The Coming of Post-industrial Society: A Venture in Social Forecasting*. New York: Basic Book. 1973.

Bottero, Wendy: "Intersubjectivity and Bourdieusian approaches to 'identity'". *Cultural Sociology*, 4(1), 2010, pp. 3–22.

Bourdieu Pierre: *The Logic of Practice*. Stanford, CA: Stanford University Press. 1990.

Bühlmann, Felix / Thomas David / André Mach: "Cosmopolitan Capital and the Internationalization of the Field of Business Elites: Evidence from the Swiss Case". *Cultural Sociology* 7(2), 2011, pp. 211–229.

Castells, Manuel: *The End of Millennium*. Oxford: Blackwell. 2000.

Delanty, Gerard: *Modernity and postmodernity : Knowledge, power and the self*. London; Thousand Oaks ; New Delhi : Sage. 2000.

Dreher, Axel: "Does Globalization Affect Growth?" *Empirical Evidence from a new Index, Applied Economics* 38(10), 2006, pp. 1091–1110.

European Commission, Brussels: Standard Eurobarometer 82, retrieved 4.1.2015, from http://ec.europa.eu/public_opinion/archives/eb/eb82/eb82_anx_en.pdf.

European Commission, Brussels: Eurobarometer Special Surveys EB 396 2013, retrieved 4.1.2015, from http://ec.europa.eu/public_opinion/archives/ebs/ebs_396_en.pdf.

Elder-Vas, Dave: "Reconciling Archer and Bourdieu in an Emergentist Theory of Action". *Sociological Theory*, 25 (4), 2007, pp. 325–348.

European Values Study 2008, 4th wave, Integrated Dataset. GESIS Data Archive, Cologne, Germany, ZA4800 Data File Version 2.0.0 2010-11-30, doi:10.4232/1.10188.

Fligstein Neil: *Euroclash: The EU, European Identity and the Future of Europe,* Oxford, Oxford University Press. 2008.

Fligstein Neil / Doug McAdam: *A Theory of Fields.* Oxford: Oxford University Press. 2012.

Fouron, Georges E. / Nina Glick Schiller: "The generation of identity: Redefining the second generation within a transnational social field". In Hector Cordero-Guzman, Robert Smith, Ramón Grosfoguel (eds.): *Migration, transnationalization and race in changing.* New York. Philadelphia: Temple University Press, 2001, pp. 292–403.

Golob, Tea: "Sodobne identifikacije v primežu transnacionalnih tokov: transnacionalna družbena polja in identifikacije "Ryanair generacije" irskih migrantov". *Dve domovini,* 40, 2014, pp. 111–122.

Golob, Tea / Sebastijan Kristovič / Matej Makarovič: "European transnational social fields and identifications: the significance of transnational participation and structured positions in the individual and national context". *European Journal of Science and Theology,* 10(5), 2014, pp. 1–20.

Hasan, Raquaiya: *Semantic Variation: Meaning in Society and in Sociolinguistics. The Collected Works of Ruqaiya Hasan, Vol 2.* London: Equinox. 2009.

Human Development Report: 20th Anniversary Edition, The Real Wealth of Nations: Pathways to Human Development. Published for the United Nations Development. 2010, retrieved 2.12.2014, from http://hdr.undp.org/en/content/human-development-report-2014.

KOF index of globalisation 2011, retrieved 4.1.2015, from http://globalization.kof.ethz.ch/query/.

Levitt, Peggy / Nina Glick Schiller: Conceptualizing simultaneity: Transnational social field perspective. *International Migration Review* 38(145), 2004, pp. 595–629.

Mau Steffen / Jan Mewes: "Horizontal Europeanisation in Contextual Perspective: What Drives Cross-border Interactions within the European Union?" *European Societies* 14, 2012, pp. 7–34.

Mingers John / Alistair Mutch / Leslie Willcocks: "Introduction: Critical Realism in Information Systems Research". *MIS Quarterly,* 37 (3), 2013, pp. 795–802.

Mishler William / Pollack Detlef: "On Culture, Thick and Thin: Toward a Neo-Cultural Synthesis". In: Pollack Detlef / Jorg Jacobs (eds) *Political Culture in Post-communist Europe: Attitudes in New Democracies.* Hants: Ashgate Publishing Company, 2003, pp. 237–257.

Morozov, Evgeny: *To save everything, Click here: Technology, Solutionism, and the Urge to Fix Problems that Don't Exist.* Penguin Group. 2013.

Mouzelis Nicos: "Habitus and Reflexivity: Restructuring Bourdieu's Theory of Practice". *Sociological Research Online* 12(6)9, 2007.

Ragin, Charles C: *Fuzzy-set Social science*. University of Chicago press. Chicago and London. 2000.

Vertovec, Steven / Robin Cohen (eds.) (2002) *Conceiving cosmopolitanism: theory, context and practice*. Oxford: Oxford University Press. 2002.

Webster, Frank: *Theories of Information Societies*. Routledge. Oxon and New York. 2014.

Jana Krivec
School of Advanced Social Studies in Nova Gorica
Gregorčičeva 19, 5000 Nova Gorica, Slovenia
jana.krivec@fuds.si

Cognitive Processes and Information Technology in Education

Abstract: *Intelligent Tutoring Systems (ITS) represent a rapidly developing intersection field between AI and cognitive science. As educational computer systems they help to improve the efficiency of the established group teaching in schools by using one-to-one teaching with instructions and feedback tailored to the individual student characteristics: his prior knowledge level, cognitive style, learning pace, etc. Although intelligent tutoring systems have already proven their effectiveness in many ITS applications, there is still room for improvement. Lately, special focus has been given on building a cognitive model that best replicates the way the human brain process information. This issue is addressed in our research which focuses on the understanding of procedural knowledge processing. More precisely, the goal of our research was to identify chunks, units of meaningful information, in procedural knowledge and to describe their characteristics. For this purpose, chess variants reconstruction analysis was used. The chess game is an appropriate research domain, since it offers several advantages:it is a complex, yet structured environment, the results obtained can be generalized to other domains, it has an objective scale for measuring strength/knowledge of chess players. On the basis of the analysis, different types of procedural chunks were described and a general model of procedural information processing was developed.These results help us to better understand human cognitive processing of procedural tasks. The knowledge acquired can be of great help for the development of the general ITS model for ill-defined problems. It could also lead to an improved interaction with ITS users – provide them with better instructions in terms of their length, context and prior knowledge consideration. Consequently, such Intelligent Tutoring Systems would be able to educate more accurately and more effectively.*

Keywords: *intelligent tutoring systems, information processing, procedural knowledge, chunks, cognitive modelling*

1. Introduction

Information technology influences our lives on every corner, even in fields that used to be reserved for humans until not long ago, e.g. teaching and giving instructions. One may argue that a human teacher could never be equivalently replaced with computer based teaching process. With the Intelligent Tutoring System (ITS) this seems to be a matter of time.

Generally, school teaching is performed in groups, not individually, even though one-to-one teaching has been proved to be far more efficient than group teaching as it can be adapted to the individual student, his prior level of knowledge, cognitive style, learning tempo, etc. Researches show that students with one-to-one instructions can master the teaching material a third time quicker than with group instructions[1]. ITS is a computer system that aims to provide not only one-to-one teaching, but also customized instructions or feedback to learners, usually without intervention from a human teacher[2]. ITS modelling is a fast growing field of cognitive and computer science, especially artificial intelligence. Many practical ITS applications have already been used, showing good results in comparison with traditional way of teaching[3 4]. The most successful is Carnegie Learning's Cognitive Tutor for mathematics, which is used by over 1300 school districts and by hundreds of thousands of students each year[5]. Students using the ITS have demonstrated larger gain in effect size over paper and pencil homework on both standardized and more conceptual, experimenter-designed tests (VanLehn et al. 2005, pp. 147–204). Although intelligent tutoring systems have proven their effectiveness, they are still not widely spread mainly because of their high development costs[6]. Lately special focus has been given on two things: building ITS that can be modelled by broad users, not necessarily programmers and building a cognitive model, that bests replicates the traits of the human information processing. The latter problem is addressed in our research which

1 Corbett, Albert T.: "Cognitive Computer Tutors: Solving the Two-Sigma Problem". In: Bauer, Mathias/ Gmytrasiewicz, Piotr J. and Vassileva, Julita (eds.): Proceedings of the 8th International Conference on User Modeling 2001 (UM 2001). Springer-Verlag: Berlin 2001, pp. 137–147.

2 Psotka, Joseph / Mutter, Sharon A.: Intelligent Tutoring Systems: Lessons Learned. Lawrence Erlbaum Associates 1988.

3 Vanlehn, Kurt et al.: "The Andes Physics Tutoring System: Lessons Learned". *International Journal of Artificial Intelligence,* 15(3), 2005, pp. 147–204.

4 Chou, Chih-Yueh/ Huang, Bau-Hung / Lin, Chi-Jen: "Complementary machine intelligence and human intelligence in virtual teaching assistant for tutoring program tracing". *Computers & Education* 57(4), 2011, pp. 2303–2312.

5 Blessing, Stephen B. et al.: "Authoring model-tracing cognitive tutors.". International Journal of Artificial Intelligence in Education, 19(2), 2009, pp. 189–210.

6 Mitrovic, Antonija et al.: »ASPIRE: An authoring system and deployment environment for constraint-based tutors.«. *International Journal of Artificial Intelligence in Education,* 19, 2009, pp. 155–188.

focuses on understanding of processing of procedural knowledge. One of the most well-known and accepted constructs of cognitive architecture are chunks – pieces of meaningful information, but the nature of chunks still remains very elusive, especially for chunks in procedural knowledge. In our research chunks in human's procedural knowledge were identified and their characteristics were described based on the reconstruction of the chess variants. Moreover, types of procedural chunks were defined and a general model of their processing was developed. The main contributions of the proposed model are the precise definition of the quantitative limitation of the chunks in our memory storage and process enhanced with the recombination mechanism of the elements in short term memory. Knowledge deriving from our research can be useful when making a cognitive model for ITS, not only for chess, but also for other general problems. Knowing how humans process information can be beneficial for defining the appropriate length and type of the instructions in different contexts and for different students preknowledge.

This chapter has the following structure: In Section 1 the current state of the ITS approaches are described. In Section 2 the theories of human cognitive architecture and chunking mechanism are presented. Section 3 addresses their shortcomings and shows improvements made by our research. Section 4 presents the main results of our research which refer to procedural chunks identification and description of their characteristics and manifestation during information processing. Models of procedural information storage and retrieve are developed. Last but not least, in Section 5 the scientific and practical contributions to understanding the human cognitive architecture and to ITS development are presented.

2. ITS approaches

In general, there are three types of ITS (cognitive tutors, constraint-based tutors, and example-tracing tutors), which possess different degrees of machine intelligence and provide different interaction mechanisms and different implementation complexities[7].

7 Chu, Yian-Shu et al.: "Implementation of a Model-Tracing-Based Learning Diagnosis System to Promote Elementary Students' Learning in Mathematics." *Educational Technology & Society,* 17(2), 2014, pp. 347–357.

2.1 Constraint-based ITS

Constraint-based ITS modelling (CBM) is based on Ohlsson's theory of learning from performance errors[8]. CBM focuses on faulty knowledge, realizing that it is not sufficient to describe what the student knows correctly. The basic assumption is that diagnostic information is not hidden in the sequence of student's actions, but in the situation (or the problem state) that the student has arrived at. The student model does not represent student's actions, but the effects of his or her actions instead[9]. In such a way, CBM represents domain knowledge as a set of state constraints. Constraints define sets of equivalent problem states. Correct problem solutions do not violate any of the constraints. Wrong answers which violate relevant constraints trigger relevant instructions. An equivalence class triggers the same instructional action. The constraint-based paradigm is feasible only for domains in which the solution itself is rich in information.

2.2 Example tracing ITS

Example tracing tutors (ETT) evaluate student behavior by flexibly comparing it against generalized examples of problem solving behavior[10]. They go beyond VanLehn's (2005) minimum criterion for ITS status, namely, that the system has an inner loop (i.e., provides within-problem guidance, not just end-of-problem feedback). Compared to other approaches ETT implements a simpler method for ITS development (no machine learning is used) that is currently more pragmatic and proven for widespread, real-world use by non-programmers. The disadvantage of these types of tutors is that they do not support tutoring across a range of similar problems, but only specific problems.

---

8 Ohlsson, Stellan: "Learning from Performance Errors." *Psychological Review*, 103(2), 1996, pp. 241–262.

9 Mitrovic, Antonija et al.: "Constraint-Based Tutors: A Success Story." In: Monostori, Laszlo / Váncza, József & Ali Moonis (eds.): Proceedings of the 14th International conference on Industrial and engineering applications of artificial intelligence and expert systems: engineering of intelligent systems (IEA/AIE '01). Springer-Verlag: London 2001, pp. 931–940.

10 Aleven, Vincent et al.: "A new paradigm for intelligent tutoring systems: Example Tracing tutors.". International Journal of Artificial Intelligence in Education, 19, 2009, pp. 105–154.

2.3 Model tracing ITS

Model tracing cognitive tutors are still among the most frequently used and successful ([11]; VanLehn et al. 2005, pp. 147–204). They apply a model-tracing approach to interpret and assess student behavior with reference to a cognitive model. A Cognitive tutor is based on a usually rule-based cognitive model of student problem-solving knowledge. A cognitive model is created by writing production rules that characterize the variety of strategies and misconceptions students may acquire. Cognitive tutors are sometimes also called "Production Rule Tutors". The first Cognitive tutors were the result of a test of Anderson's Atomic Component of Thought (ACT) Theory of cognition: if cognitive skills can be realized as units of goal-related knowledge, then successful learning should be accomplished by teaching students using those units[12]. During problem solving, the Cognitive tutor follows the student in a process called model-tracing. In model-tracing, the tutoring system maintains a model of problem solving that is 'traced' (compared) against the student's actions. Intelligent tutoring systems consist of four basic components based on a general consensus amongst researchers[13]: The Domain model, The Student model, The Tutoring model, and The User interface model. Feedback during problem-solving is given based on the current state of the model (also called working memory) and the rules that represent student cognition and action. The cognitive model is commonly applicable to various problems in the same domain. For this type of ITS to function well a good cognitive model is needed. "For example, the Andes physics tutor by VanLehn and his colleagues (2005) contains a cognitive model of how to solve problems within physics. This model is referenced for each step in the problem solving process in order to make sure the student stays on path. The model can also be used to provide hints and other assistance to the student learning the material" (Blessing et al., 2009, pp. 189–210). Making a good cognitive model is one of the toughest tasks when developing ITS. Numerous cognitive theories, cognitive models and cognitive models assistant programs were developed. Chu et al. (2014) developed a Model-tracing Intelligent Tutor (MIT) to

---

11 Ritter, Steven et al.: "What evidence matters? A randomized field trial of Cognitive Tutor Algebra I.". In: Hirashima, Tsukasa / Hoppe Ulrich / Young Shelley Shu (eds.), Supporting Learning Flow through Integrative Technologies, 162. Amsterdam: IOS Press 2003, pp. 13–20.

12 Anderson, John R. et al.: "Cognitive tutors: Lessons learned.". The Journal of Learning Sciences, 4, 1995, pp. 167–207.

13 Nkambou, Roger / Mizoguchi, Riichiro / Bourdeau, Jacqueline: *Advances in intelligent tutoring systems.* Heidelberg: Springer 2010.

diagnose students' learning problems and provide learning feedback for individual students, Aleven et al. (2009) developed Cognitive Tutor Authoring Tools (CTAT), to name a few. Despite that, there is still a lot of space for improvements and open issues to overcome in this field. This makes this area one of the highlights in current artificial intelligence (AI) and psychology research.

In the following sections we focus on theories of human cognitive architecture and functioning which are essential for developing a good cognitive model for ITS. Special emphasis is on the presentation of the chunking mechanism, to which our research contributes new findings and better understanding. Last but not least, further challenges in the field of ITS from psychological and AI perspectives are discussed.

3. Human cognitive architecture and functioning

When giving instructions, in order to provide the right information at the right time, one must know how real human brains store and retrieve i.e. process appropriate information. The basic model of human information processing is the Atkison –Shiffrins model of memory. The model asserts that human memory has three separate components[14]:

1. a sensory register, where sensory information enters memory. With the help of the law of attention and later prior knowledge a person first notices the relevant information.
2. a short-term memory, also called working memory (WM), which receives and holds input from both the sensory register and the long-term store, and
3. a long-term store, where information which has been rehearsed (explained below) in the short-term memory is held indefinitely. When needed, information is recalled from LTM using appropriate clues.

According to the descriptions of basic perceptions, mechanisms of learning and recall from memory and used methodology, the theories are divided into three main branches: perceptual, knowledge-based and process - i.e. goal-oriented theory. The first focuses on the interpretation of the storage structures (as Chase and

14 Atkinson, Richard / Shiffrin, Richard: Chapter: Human memory: A proposed system and its control processes. In: Spence, Kenneth Wartenbee / Spence, Janet: The psychology of learning and motivation (Volume 2). New York: Academic Press 1968. pp. 89–195.

Simon's Chunking theory[15], Gobet and Simon's Template theory[16]). Knowledge-based paradigm interprets substantively different organization of information in relation to knowledge (Ericsson and Kintsch theory of long-term working memory – LTWM[17]). Goal-oriented theories explain information processing and pattern recognition as a conscious process that leads to the execution of an action or a certain goal. Such theories are Andersson and Lebiers Atomic component of the Thought – Rational (ACT – R) theory[18], the SOAR (*State operator and Result*) model[19], ICARUS architecture[20] and the EPIC (*Executive Process – Interactive Control*) model[21].

Most of the existing theories of human information processing (Chunking theory, Template theory and ACT theory) have one concept in common - chunks. Chunks are pieces of meaningful information (*i.e.,* information that has a relation to information in long term memory) which enable easier learning and perception[22]. ITS constructors use the chunking mechanism mostly in declarative part of the cognitive structure, but rarely in the procedural part, which is most of the time defined with if-then rules (but not chunks of rules). We might consider the number of nodes in the rule as some kind of chunks. However, chunks might represent the number of rules (instructions) that can be shown to a student at the same time. If we have clear definition of chunks in declarative and procedural knowledge it would be of a great help when constructing ITS. It may help in defining how much information to include in one step instruction and how many steps are the optimal amount that students with different prior knowledge can

---

15 Chase, William / Simon, Herbert: »The mind's eye in chess.« In: Chase William (Ed.) *Visual information processing.* New York: Academic Press, 1973, pp. 215–281.

16 Gobet, Fernand / Simon, Herbert A.: »Templates in chess memory: A mechanism for recalling several boards.«. Cognitive Psychology, 31, 1996, pp. 1–40.

17 Ericsson, K. Anders / Kintsch, Walter: »Long-term working memory.« *Psychological Review,* 102 (2), 1995, pp. 211–245.

18 Anderson, John Robert / Lebiere, Christian: *The atomic components of thought.* Hillsdale, NJ: Lawrence Erlbaum Associates 1998.

19 Newell, Allen: *Unified Theories of Cognition.* Cambridge, MA: Harvard University Press 1990.

20 Langley, Pat / Choi, Dongkyu: »Learning recursive control programs from problem solving.«. *Journal of Machine Learning Research,* 7, 2006, pp. 493–518.

21 Kieras, David / Meyer, David E.: »An overview of the EPIC architecture for cognition and performance with application to human-computer interaction.«.*Human Computer Interaction,* 12, 1997, pp. 391–438.

22 Miller, George A.: "The Magical Number Seven, Plus or Minus Two: Some Limits on our Capacity for Processing Information." *Psychological Review,* 63, 1956, pp. 81–97.

still process and store. However, there is neither a general definition of size and number of chunks in human information processing, nor a described mechanism for the procedural chunks storing and recalling, especially for the procedural type of information. In the following section we present the known facts about chunking, the findings of our research and a model, developed for the explanation of chunks processing.

3.1 Chunking

Although the concept of chunks is widespread and universally accepted, the nature of chunks is far from being clearly defined and understood. Miller can be considered as the "father of chunks". He first explained the possibility of increasing the efficiency of working memory (WM) by using a simple mental recoding of information into a smaller number of more complex related units, called "chunking". "Is the capacity of WM depended on the size, number of chunks or none of the above?" is the point of conflict of many authors since Miller`s famous 7±2 magical number, representing the capacity of WM. We will address this further in this chapter. Simon and Chase[23] generally applied the concept of chunks to long-term memory, defining chunks as units of perception and meaning, while chunking can be seen as a way of storing information in long-term memory. Chunks are best described in the Gobet and Simons Template theory. They are defined as elements in a discrimination network that are joined together closer than they are joined with other elements in the network. They are further developed into templates, chunk networks which can have some elements that are not fixed and can be changed from one chunk to another.

According to Anderson (1995), there are two essential components of stored information: declarative knowledge and procedural knowledge. Declarative knowledge collects the factual or conceptual knowledge that a person possesses. The declarative chunks were being explored into details such as size of chunks; some authors argued that their size is three elements[24], while the newest researches support the idea of bigger chunks: from 4±2 elements (Gobet and Simon, 1996)

---

23 Simon, Herbert / Chase, William: "Perception in chess.". Cognitive Psychology, 4, 1973, pp. 55–81.

24 Allen, Philip A. / Coyne, Andrew C.: "Age differences in primary organization or processing variability? Part II: Evidence for processing variability." *Experimental Aging Research*, 14, 1988, pp. 151–157.

up to 10–15 elements as states the Template theory. It was also discovered, that with age chunks are shrinking and bigger problems occur at chunk recalling[25].

3.2 Chunks in procedural knowledge

According to Sowa's opinion, the paradigm of declarative knowledge construction has largely failed to produce human-like cognitive processing in computers[26]. Self[27] showed that focusing on the process by which knowledge is constructed is more important than focusing on the target knowledge itself. By using the descriptions of classes, properties, instances and the descriptions of their relationships in ontology, the system can provide a more robust function on reasoning. Most noticeable cognitive theories that try to explain procedural knowledge are:

1. »Flowchart model« – which explains procedural knowledge in a sense of mental flowcharts, that defines the steps of a particular process applied to the information that enters someone's cognitive system[28].
2. Von Neumann's model – is based on computational processing. It assumes that information processing includes consecutive operations, guided by the rules of the symbolic representations (Wyer and Srull, 1994). In this framework Lu[29] stated that in designing a generic architecture to represent procedural knowledge, the actions defined by domain experts and the control of action flow are two important tasks. Thus this model is the combination of behavior model (procedural knowledge) and ontology (declarative knowledge). In the so-called InfoMap, they proposed an ontological representation scheme called the Process Map (PM) to represent procedural knowledge. The Process Map (PM) is a way to represent procedural knowledge, which can be treated as a series of processes connected by junctions and links. The direction of the flow of each instance is decided by the preconditions of each process. The

25 Cowan, Nelson / Chen, Zhijian / Rouder, Jeffrey N.: "Constant capacity in an immediate serial-recall task: A logical sequel to Miller." Psychological Science, 15, 2004, pp. 634–640.

26 Sowa, John F.: *Knowledge Representation: Logical, Philosophical, and Computational Foundations.* Pacific Grove, CA, USA: Brooks Cole Publishing 1999.

27 Self, John. »The defining characteristics of intelligent tutoring systems research: ITSs care, precisely.«. *International Journal of Artificial Intelligence in Education*, 10, 1999, pp. 350–364.

28 Wyer, Robert S. Jr. / Srull, Thomas: *Handbook of social cognition: Vol. 1, Basic processes (2nd ed.).* Hilsdale, NJ: Erlbaum 1994, pp. 1–40.

29 Lu, Chun-Hung et al.: "Ontological Support in Modeling Learners' Problem Solving Process." *Educational Technology & Society,* 8(4), 2005, pp. 64–74.

steps of problem solving may also be recorded, as long as the processes are clearly defined. This is useful for people who want to detect the errors or track the history of a procedure. By adopting the InfoMap, authors hope to help teachers construct curricula (declarative knowledge) and teaching strategies by capturing students' problem-solving processes (procedural knowledge) dynamically. Using the concept of declarative and procedural knowledge in intelligent tutoring systems, we can accumulate and duplicate the knowledge of the curriculum manager and student (Lu at al., 2005).

3. Processing models – assume that procedural knowledge is changed during its usage. The more one action is used the more it is automated, which represents procedural knowledge. The model offers a declarative description of procedural knowledge. It is used in ACT-R cognitive architecture and Logan's model of memory based automation[30].
4. »Parallel distributed processing models« have their origins in neural networks information processing. It is functioning in a distributed and parallel way rather than consecutive. It assumes that knowledge is not stored explicitly, but in elements interconnection. Learning enhances particular connections. The most famous parallel distributed model is proposed by Cohen[31].

As for the chunks, some modern researches are proving the existence of chunks in the procedural knowledge[32 33]. Chunks identification was mostly based on the longer time between a recall of two consequent elements in the recall task or on the longer pause between the writing of two consequent elements. Other proofs are based on the more successful recall of information when using chunks[34].

---

30 Logan, Gordon D.: "Toward an Instant Theory of Automatization." *Pyschological Review*, 95, 1988, pp. 492–527.

31 Cohen, Jonathan D. / Dunbar, Kevin / McClelland, James: "On the control of automatic processes: A parallel distributed processing account of the Stroop effect." *Psychological Review*, 97, 1990, pp. 332–361.

32 Jiménez, Luis: "Taking patterns for chunks: Is there any evidence of chunk learning in continuous serial reaction-time tasks?". *Psychological Research*, 72, 2008, pp. 387–396.

33 Cheng, Peter C-H. C / Rojas-Anaya, Hector (2005). »Writing out a temporal signal of chunks: patterns of pauses reflect the induced structure of written number sequences.«. In: Bara, Bruno G., Barsalou, Lawrence / Bucciarelli, Monica (ed.), *Proceedings of the Twenty Seventh Annual Conference of the Cognitive Science Society*. Mahwah, NJ: Lawrence Erlbaum 2005, pp. 424–429.

34 Bapi, Raju S. et al.: »Investigation of sequence processing: A cognitive and computational neuroscience perspective.« *Current Science*, 89(10), 2005, pp. 1690–1698.

However, there are several shortcomings of the presented theories when trying to identify concrete chunks or explain the nature and mechanism of procedural chunks.

4. Shortcomings of the present theories and researches on procedural chunking and improvements made by our research

Shortcomings of the described researches primarily refer to methodological procedures used. For example, Chase and Simon (1973) differentiate between two chunks by the time interval in recall which is longer than two seconds. However, there are big individual differences among people in information processing tempo.

Furthermore, most of the researches have the next shortcomings:

- Many researches do not include enough participants, especially experts. In Chase and Simon's study there were three experts chess players, which is far less than the minimal research standards in academicals. Moreover, researches usually did not included world-class experts in the actual field of the research.
- None of them includes more than one indicator to identify chunks, possibly yielding chunks that can be less accurate and less objective.
- Researches did not divide participants into groups according to their level of expertise in the relevant field. This could blur the chunk borders because they join data of experts, which may have bigger chunks and data of amateurs which may be smaller – the final chunks can therefore be of medium size.
- All the studies are also missing the findings about the size and number of recalled chunks in conjunction with the context. Namely, according to the Cognitive Load Theory[35] the context is proved to be very influential when processing information. Consequently, no context influence on chunks size or recall number is provided in the studies so far.
- There is also no research that deals with the influence of the sort order of the elements in chunks on recall. We addressed these issues in our research about procedural chunks characteristics and processing.

In our research we avoided the shortcomings of the previous researches. The chess game was used as the research domain, since it offers several advantages: It is a complex yet structured environment; results can be generalized to other domains; it has an objective scale for measuring strength/knowledge of the chess players.

---

35 Anderson, John Robert / Schooler, Lael: "Reflections of the environment in memory." *Psychological Science*, 2, 1991, pp. 396–408.

We examined the assumptions with legal and illegal chess variants reconstruction tasks (32 movements of pieces). With the use of standardized time interval information, the individual differences in the reconstruction tempo were overcome. The identification of chunks was done with adaptations of the existing methods and developing new methods. We combined the results of the methods that take into account different criteria for chunks determination, namely: time, collective performance, association rules, clustering and key positions. This way we avoided potential mistakes when using just one method of chunk extractions which could be influenced by biased factors. Furthermore, we tested chess players with different chess knowledge (expressed by the ELO rating – ELO rating system is a method for calculating the relative skill levels of players in competitor-versus-competitor games such aschess) – from top world grandmasters to amateurs. Next, we included the control group showing that it was not a better memory that caused better recall of the chess moves, but the knowledge connected to the presented moves, which made the recall easier and better. For the detailed description of the methods used please see the article Krivec et al.[36].

5. Procedural chunks identification, their characteristics and manifestation during information processing

The results of our research have initially shown that the performance (efficiency and speed) of the procedural information processing relates to the level of relevant background knowledge and, to a lesser extent, gender and age. Consequently, chunks in procedural knowledge were identified separately for each group of participants. We found that chunks are built of smaller units. Based on the observations we have developed models of types and the possible manifestations of procedural chunks in information processing, which vary according to the flexibility of the elements in the chunk and chunks semantic homogeneity. In the following section these models are presented in details.

36 Krivec, Jana / Guid, Matej / Bratko, Ivan: "Identification and characteristic descriptions of procedural chunks." In Dini, Petre (Ed.). *Proceedings of Computation world 2009, Computation world: future computing, service computation, adaptive, content, cognitive, patterns.* New York: IEEE Coputer Society, 2009, pp. 448–453.

5.1 Models, Types and Manifestation of Chunks in Complex Procedural Knowledge, Defined with Time Sequential Order

Information in long term memory is stored in different ways:

- Sort order of the elements in chunks can be fixed or not.
- Chunks may consist of fixed elements only or can include some variable/not determined elements.
- Chunks may consist of one or more substantive (sub)chunks.

We proposed several different types of chunks and their manifestation in complex procedural knowledge recall, according to the sort order of the elements in the chunk.

»*Simple*« chunk consists of one substantive chunk with a defined sort order.

Figure 1: Graphic illustration of a "Simple chunk".

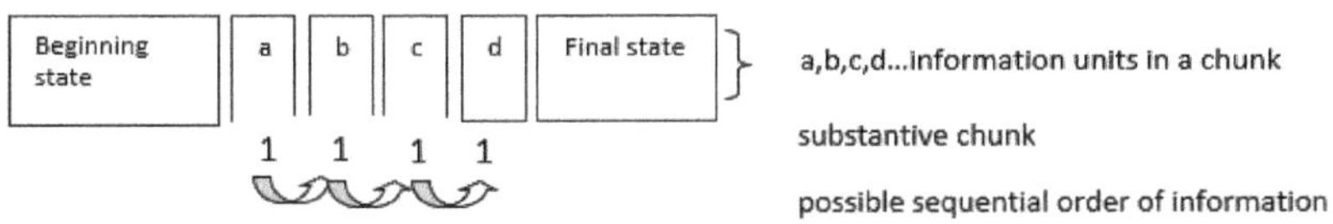

The simple chunk is the "strongest" type (most frequently used), most internally connected and cannot be recombined (changed) during recall. They have advantage at recall, so called higher activation power (ACT theory) which demands less energy for information processing.

»*Mixed*« chunk: consists of one substantive chunk that includes partial or fully flexible sort order of its elements while the beginning and the final end states of the procedural chunk are fixed. The proposition that a chunk can be »mixed« implies that elements in such a chunk can be replaced with any other element or that sort order of the elements in the chunk can be irrelevant.

Figure 2: Graphic illustration of a "Mixed chunk".

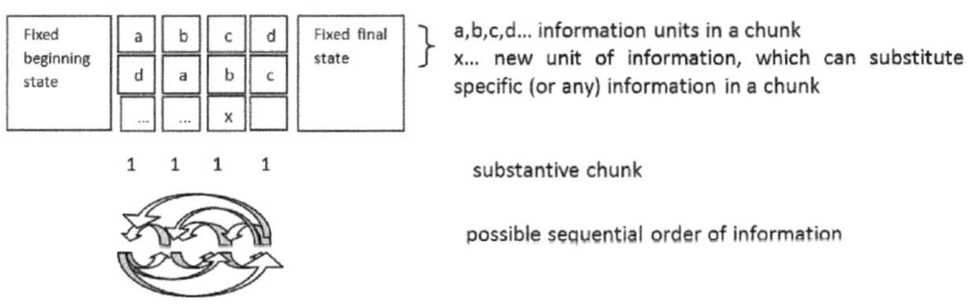

»Covering« chunk: two or more substantive chunks with a fixed sequential order join, because they are usually interrelated, formulating a bigger latent chunk.

Figure 3: Graphic illustration of a "Covering chunk".

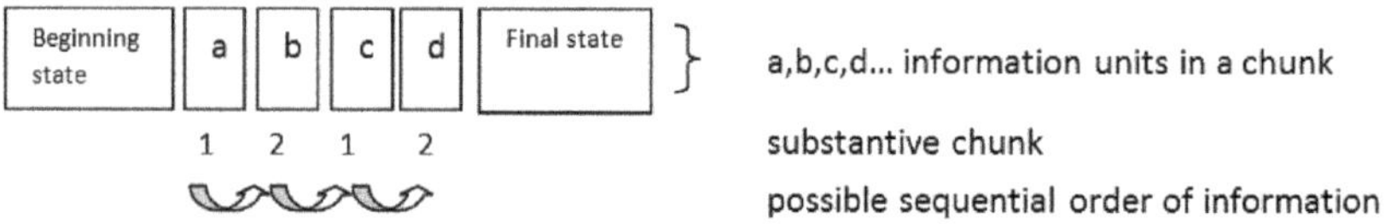

Figure 4. Graphic illustration of a "Covering chunk".

»Covering mixed« chunk: two or more substantive chunks that include a partly fully flexible sequential order and have a fixed beginning and final state.

Figure 4: Graphic illustration of a "Covering mixed chunk".

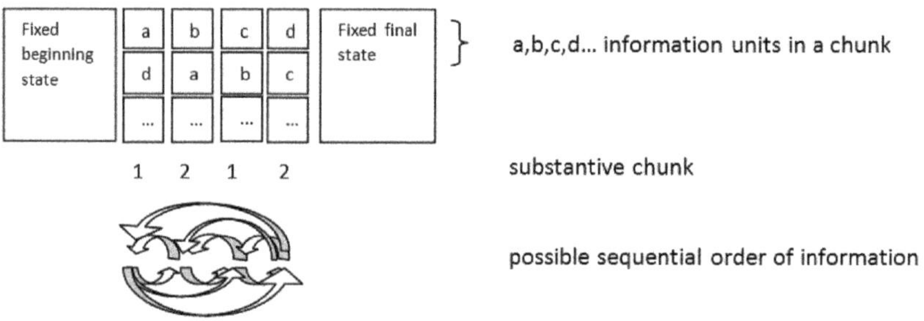

With regard to the size of the chunk many authors concluded that capacity of each chunk amounts to three elements (Allen and Coyne, 1988). Chase and Simon (1973) and Wickelgren[37] showed that this capacity is much too small. Template theory (Gobet and Simon, 1996) developed a new term – templates, which are big chunks (up to 32 elements, but usually between 10 and 15 elements) that have some fixed and some variable elements. A greater number of elements in a chunk were later confirmed in the research of Cooke et al.[38]

We found that the size of the procedural chunks and its retrieval speed varies among people with different background knowledge. People with more background

37 Wickelgren, Wayne A.: »Single-trace fragility theory of memory dynamics.« *Memory & Cognition*, 2, 1974, pp. 775–780.

38 Cooke, Nancy J. et al.: »Role of high-level knowledge in memory for chess positions.« *American Journal of Psychology*, 106, 1993, pp. 321–351.

knowledge operate on average faster and with larger basic chunks (four units) than those with less prior knowledge (three units), while there is no such difference between the groups when considering detailed (sub)chunks (2 to 2.5 units).

Our results further showed that experts' knowledge is semantically not organized differently from non-experts' knowledge, as presumed by knowledge-based theories. It rather seems that experts combine smaller (detailed) chunks into more complex, larger (basic) knowledge units.

Based on the results we further assume, that the limited capacity of working memory refers to the limited amount of material (mostly a combination of numbers and size of chunks), which can be simultaneously retained in working memory. The results also indicated that there is a difference between the amount of retrieved information and the amount of information that can be actively processed in the short term memory.

What the capacity of short term memory is, has been a matter of debate for a long time[39]. Does it depend on the number of recalled chunks as stated by Cowan et al. (2004), on the size of the chunks recalled as stated by Baddeley et al.[40] or on a combination of both? Miller (1956) made a first assumption that 7±2 elements can be simultaneously recalled to short term memory (STM). Later, Template theory (Gobet and Simon, 1996) reduced this amount to only three elements. Cowan[41] stated that the capacity of STM is 3–5 chunks. MacGregor[42] on the other hand showed that the capacity of STM is a combination of the size and the number of chunks. He showed that for a 6-element chunk the optimal recall strategy is three 2-elements chunks, two 3-elements chunks or one 6-elements chunk.

We showed that people with more prior knowledge often simultaneously trigger seven chunks to STM, where they can actively operate with only around four chunks. This ratio is four: 2–3 for people with less background knowledge. Based on the results of the study and our observations, we proposed a procedural information processing model. Its main novelty in comparison with previous models is the recombination module, where two or more procedural chunks that were

---

39 Chen, Zhijian / Cowan, Nelson: »Chunk limits and length limits in immediate recall: A reconciliation.« *Journal of Experimental Psychology: Learning, Memory, and Cognition*, 31, 2005, pp. 1235–1249.

40 Baddeley, Alan David / Thomson, Neil / Buchanan, Mary: »Word length and the structure of short-term memory.« *Journal of Verbal Learning and Verbal Behavior*, 14, 1975, pp. 575–589.

41 Cowan, Nelson: *Working memory capacity.* New York, NY: Psychology 2005.

42 MacGregor, James N.: »Short-term memory capacity: Limitation or optimization?« *Psychological Review*, 94(1), 1987, pp. 107–108.

recalled into working memory recombine in order to fulfill the semantic and context request of the perception task.

5.2 Model of Sequential Information Organization and Processing

We propose the following model of organization and processing of sequential info rmation.

Figure 5: Model of processing and storing of complex information of procedural type.

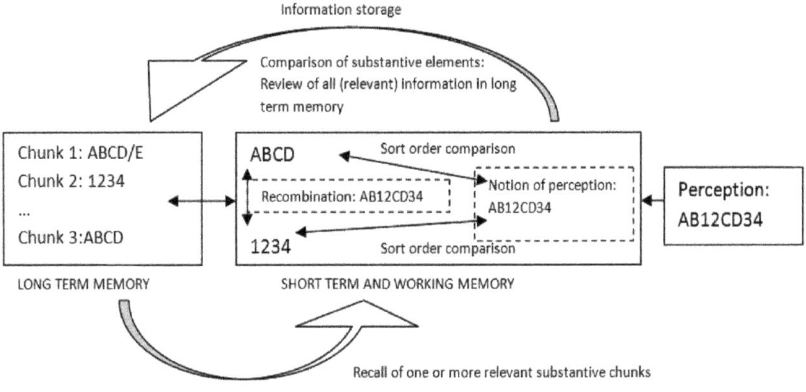

The model represents the flowchart of the information flow among different memory units. It can be seen that the processing and the storing/organizing of information are interconnected among each other. The model does not presuppose separate units for chunks and sort order. Chunks are notions of information in long term memory. Sort order is a procedure which is carried out on all chunks recalled as verification of matching with perception notion in working memory. Our model suggests that people are capable of simultaneous processing of more than one sequence of information. Namely, simple sequential recall has problems with explanation of the recall failures[43]. Therefore, it is not always about simple sequential recall of information chunks, but also about parallel recall and manipulation of information chunks. Further on, we describe these separate processes in detail.

43 Henson, Richard: »Short-term memory for serial order: the start-end model.« *Cogntive Psychology*, 36, 1998, pp. 73–137.

Information storage

Our proposition is that information can be stored in different kinds of chunks depending on whether the stored information is fixed or variable (loose). In the latter case, chunks can be changed by combining two smaller units into a one bigger, more complex unit or by converting "loose" chunks (with frequent usage) to more stable chunks with fixed elements.

Processing of information comparison and recall

Our model proposes the following way of comparison and recall of procedural information: firstly, the information perceived by our sensory system is temporarily retained in our working memory. Elements of the retained sensory information chunks ("original elements") are than compared with elements of chunks stored in our long term memory. It lasts until all the substantive original elements are recalled. This does not necessarily mean that one chunk of stored information covers all the original elements. More LTM chunks can be recalled at once. The size of chunks varies according to the amount of relevant prior knowledge (around 4±2 units of information). Regardless of the amount of knowledge the upper limit of recalled chunks that can be retained in STM is 7±2. Chunks can possibly include data about sort order. When all of the original elements are covered, the process of selected chunks recall is activated. During this process, the rule of the least possible chunks that still cover all the elements is always followed. This is possible due to different kinds of stored chunks that have different activation power. Firstly, chunks with fixed elements and their sort order are recalled, followed by the chunks that allow variable elements or sort order. When chunks are recalled from LTM their notion is retained in STM.

Closure of the information searching

Some theories propose that a fixed searching interval of 30s is needed for the recall of appropriate information. Others disagree and propose a flexible searching interval[44]. We propose that information searching is completed when all the elements needed are covered within the least number of stored chunks. We assume that the interval is the shortest when one chunk is enough to cover all the elements, while a longer interval is needed when more chunks are being recalled.

---

44 Dougherty, Michael R. / Harbison, J. Isaiah: Motivated to retrieved: How often are you willing to go back to the well when the well is dry? *Journal of Experimental Psychology: Learning, Memory, and Cognition*, 33, 2007, pp. 1108–1117.

The next question is whether the searching interval depends on the size of the recalled chunk(s) or not. We assume it does. Namely, in the case of bigger chunks more elements need to be compared and this is more time consuming. Working memory capacity is limited. We assume that limitation addresses the combination of the number and size of the information/chunks that are simultaneously retained in working memory.

Recombination

When chunks are recalled to STM (where their notion is retained), sort order of the elements is compared to the sort order of the original elements. If the sort orders of the recalled elements is not the same as the original elements' sort order, elements recombination is performed. Recombination can happen in one chunk (recombination of sort order) or among two or more recalled chunks (recombination of elements and sort order). The idea of recombination follows the findings on DNA sequences[45]. The main idea is that »*deletion*«, »*substitution*«, »*translocation*« and »*insertation*« which are typical for DNA, exist also in the alleged mistakes of STM: *deletion=omission (forgetting)*, »*substitution*« = confusion (one element is confused with other), »*translocation*« = movement mistake (element is recalled at wrong place) and »*insertation*« = insertion mistake (a redundant or wrong element is inserted while processing or incorrectly repeated). As previously mentioned, all chunks are not able to recombine. Only chunks with variable elements (as chunk 1 on the picture of the model) can be recombined. Fixed chunks (as chunk 3 ABCD on the picture) cannot be divided and recombined. They have become fixed because they were used in exactly the same form many times and as such allow quick and effortless recall. They may be involved in the recombination process, but only as the whole part of a bigger recombination that includes a variable chunk. Persons with less relevant domain knowledge are more likely to recall fixed chunks than chunks with variable elements, which are more likely to appear with relevant domain experts.

6. Chunking and context

Anderson and Schooler (1991) stated in their ACT-R theory that information processing efficiency is affected by the characteristics of the context. Context influences information processing through its general perception traits such as

45 Alberts, Bruce et al.: Molecular Biology of the Cell, Chapter 5: DNA Replication, Repair, and Recombination. New York, Garland Science 2002.

information dispersion, complexity and amount. The thesis was developed in the «Cognitive Load Theory« – CLT, educational theory, based on the assumption of the limited capacity of the STM, according to the amount of information simultaneously retained in STM and the amount of operation simultaneously performed on the recalled information (Anderson and Schooler, 1991). In our research we examined how the processing of information and chunking relates to different characteristics of context. Context was defined with different characteristics of chess pieces configuration and chess variants (e.g. moves dispersion, numbers of pieces on the board, number of pieces included in the variant, etc.)

Our results support the assumptions of the Cognitive load theory that an excessive burden on working memory with external factors (dispersion and complexity of information in context) may lead to less efficient information processing and to smaller chunks.

7. Contribution of our findings to ITS and its future challenges

So far, cognitive tutors have been used successfully but are still used mainly for tutoring subject areas that have an optimal balance of production rules, complexity and maximum benefit to the learner such as algebra, computer programming and geometry. However, the final goal of the ITS experts is to build a general computer-based tutor that will achieve the effects of learning individually with a human tutor on all learning topics (Mitrovic et al., 2009). Such ITS should incorporate student knowledge, preferred learning strategies, areas of interest besides that of instruction, preferred presentation style, level of concentration and so on. Consequently, cognitive modeling in the future should be precise, general and opened up to a wider class of people. Such a model would represent a base for many models of ITS for ill-defined problems such as ethics and could be implemented to all the domains with the help of a domain specific expert. Further, it would help to give better instructions to the user, help with the construction of a more interactive dialogue between the user and ITS.

We believe that our findings have made an important contribution towards achieving the above-described goal by developing an integrated understanding of human cognitive processing and expertise of procedural tasks. Our findings redefined or improved existing models of procedural memory and cognitive architectures.

In the first place, the results have shown, that the performance (efficiency and speed) of the procedural information processing relates to the level of relevant background knowledge, to a lesser extent, gender and age. Consistent chunks were identified in procedural knowledge for each group of participants. If we know what is the typical human expert chunk size and its way of processing, questions like this can be answered: How many concepts shall be presented at once, depending on the student's relevant prior knowledge and context? Consequently, ITS would provide the knowledge more accurately and more effectively.

Next, we found that the stronger (basic) chunks are built of smaller (sub) units – detailed chunks. This means that knowledge of content experts is not organized semantically differently from non experts, as presumed by knowledge-based theories. It rather seems that experts combine smaller (detailed) chunks into a more complex, larger (basic) knowledge units. Based on the results we further assumed, that the limited capacity of working memory refers to the limited amount of material (mostly a combination of numbers and size of chunks), which can be simultaneously retained in working memory. The results also indicated that there is a difference between the number of retrieved information and the number of information that can be actively processed in the short term memory. We further concluded, that people with more prior knowledge often simultaneously trigger seven chunks to short-term memory, where others can only actively operate with around four chunks. While among people with less background knowledge this ratio is four: 2–3. Based on the results of the study and our observations, we proposed procedural information processing model. Its main novelty in comparison with previous models is the recombination module, where two or more the procedural chunks recalled into working memory recombine in order to fulfill the semantic and context request of the task. Last but not least, we examined how the processing of information and chunking relates to different characteristics of context. The results support the assumptions of the Cognitive load theory that an excessive burden on working memory with external factors (dispersion and complexity of information in context) may lead to less efficient information processing and to smaller chunks.

The obtained results contribute to an integrated understanding of human cognitive processing and expertise of procedural tasks and refining existing models of procedural memory and cognitive architectures. This knowledge may have a great impact on the student cognitive modelling, instructions and ITS development.

Nevertheless, there are other challenges, not directly relevant to our research, that ITS should cope with in the near future.

The qualitative aspect of a student answer to a presented problem will have to be taken into consideration. The further explanation or problem presentation should be shaped not only based on (in)correctness of the provided answer, but also on the base of the answer content.

A lot of authors are trying to include games when teaching with ITS – so-called gaming. This can be quite challenging for complex and difficult problems – how to create a game, that would teach us specific, complex and difficult content. However, ITS is also appropriate to teach children in a form of intelligent games than makes children learn through playing, effortlessly. MacFarlane and coworkers reported that 85% of parents interviewed said that they thought their children could learn something from playing computer games[46]. They identified skill development in areas such as decision making, design, strategy, cooperation, and problem solving. Using computer games in education can enhance the motivation among students.

Lately, the aims are to develop an ITS for intelligent support in group education. This aspect should incorporate and consider interindividual differences as well as social interactions.

Last but not least, ITS are flirting with engaging metacognition and even affects in computer based learning. The next challenge would be to include neuroscience when constructing ITS.

8. References

Alberts, Bruce / Johnson, Alexander / Lewis, Julian / Raff, Martin / Roberts, Keith / Walter, Peter: *Molecular Biology of the Cell, Chapter 5: DNA Replication, Repair, and Recombination.* New York, Garland Science 2002.

Allen, Philip A. / Coyne, Andrew C.: "Age differences in primary organization or processing variability? Part II: Evidence for processing variability.". *Experimental Aging Research*, 14, 1988, pp. 151–157.

Aleven, Vincent / McLaren, Bruce M. / Sewall, Jonathan / Koedinger, Kenneth R.: "A new paradigm for intelligent tutoring systems: Exampletracing tutors.". *International Journal of Artificial Intelligence in Education*, 19, 2009, pp. 105–154.

Anderson, John Robert / Corbett, Albert T. / Koedinger, Kenneth R. / Pelletier, Ray: "Cognitive tutors: Lessons learned.". *The Journal of Learning Sciences*, 4, 1995, pp. 167–207.

Anderson, John Robert / Lebiere, Christian: *The atomic components of thought.* Hillsdale, NJ: Lawrence Erlbaum Associates 1998.

Anderson, John Robert / Schooler, Lael: "Reflections of the environment in memory.". *Psychological Science*, 2, 1991, pp. 396–408.

46 McFarlane, Angela / Sparrowhawk, Anne / Heald, Ysanne: Report on the educational use of games: *An exploration by TEEM of the contribution which games can make to the education process.* Cambridgeshire, UK 2002.

Atkinson, Richard / Shiffrin, Richard: Chapter: Human memory: A proposed system and its control processes. In: Spence, Kenneth Wartenbee / Spence, Janet: The psychology of learning and motivation (Volume 2). New York: Academic Press 1968. pp. 89–195.

Bapi, Raju S. / Pammi, V. S. Chandrasekhar / Miyapuram, Krishna P. / Ahmed, Kazuyuki Samejima: »Investigation of sequence processing: A cognitive and computational neuroscience perspective.«. *Current Science, 89(10),* 2005, pp. 1690–1698.

Baddeley, Alan David / Thomson, Neil / Buchanan, Mary: »Word length and the structure of short-term memory.«. Journal of Verbal Learning and Verbal Behavior, 14, 1975, pp. 575–589.

Blessing, Stephen B. / Gilbert, Stephen B. / Ourada, Stephen and Ritter, Steven: "Authoring model-tracing cognitive tutors.". *International Journal of Artificial Intelligence in Education*, 19(2), 2009, pp. 189–210.

Chase, William / Simon, Herbert: »The mind's eye in chess.« In: Chase William (Ed.) *Visual information processing.* New York: Academic Press, 1973, pp. 215–281.

Chen, Zhijian / Cowan, Nelson: »Chunk limits and length limits in immediate recall: A reconciliation.«. *Journal of Experimental Psychology: Learning, Memory, and Cognition, 31,* 2005, pp. 1235–1249.

Cheng, Peter / Rojas-Anaya, Hector (2005). »Writing out a temporal signal of chunks: patterns of pauses reflect the induced structure of written number sequences.«. In: Bara, Bruno G., Barsalou, Lawrence / Bucciarelli, Monica (ed.), *Proceedings of the Twenty Seventh Annual Conference of the Cognitive Science Society*. Mahwah, NJ: Lawrence Erlbaum 2005, pp. 424–429.

Chou, Chih-Yueh / Huang, Bau-Hung / Lin, Chi-Jen: "Complementary machine intelligence and human intelligence in virtual teaching assistant for tutoring program tracing". *Computers & Education*, 57(4), 2011, pp. 2303–2312.

Chu, Yian-Shu / Yang, Haw-Ching / Tseng, Shian-Shyong / Yang, Che-Ching: Implementation of a Model-Tracing-Based Learning Diagnosis System to Promote Elementary Students' Learning in Mathematics. *Educational Technology & Society*, 17(2) 2014, pp. 347–357.

Cohen, Jonathan D. / Dunbar, Kevin / McClelland, James: "On the control of automatic processes: A parallel distributed processing account of the Stroop effect.". *Psychological Review*, 97, 1990, pp. 332–361.

Cooke, Nancy J., Atlas, Robert S., Lane, David Mark / Berger, Robert C. : »Role of high-level knowledge in memory for chess positions.«. *American Journal of Psychology*, 106, 1993, pp. 321–351.

Corbett, Albert T.: "Cognitive Computer Tutors: Solving the Two-Sigma Problem". In: Bauer, Mathias/ Gmytrasiewicz, Piotr J. and Vassileva, Julita (eds.): *Proceedings of the 8th International Conference on User Modeling 2001 (UM 2001).* Springer-Verlag: Berlin 2001, pp. 137–147.

Cowan, Nelson / Chen, Zhijian / Rouder, Jeffrey N.: "Constant capacity in an immediate serial-recall task: A logical sequel to Miller.". *Psychological Science*, 15, 2004, pp. 634–640.

Cowan, Nelson: *Working memory capacity.* New York, NY: Psychology 2005.

Dougherty, Michael R. / Harbison, J. Isaiah: Motivated to retrieved: How often are you willing to go back to the well when the well is dry? *Journal of Experimental Psychology: Learning, Memory, and Cognition*, 33, 2007, pp. 1108–1117.

Ericsson, K. Anders / Kintsch, Walter: »Long-term working memory.«. *Psychological Review, 102 (2),* 1995, pp. 211–245.

Gobet, Fernand / Simon, Herbert A.: »Templates in chess memory: A mechanism for recalling several boards.«. *Cognitive Psychology*, 31, 1996, pp. 1–40.

Henson, Richard: »Short-term memory for serial order: the start-end model.«. *Cogntive Psychology*, 36, 1998, pp. 73–137.

Jiménez, Luis: "Taking patterns for chunks: Is there any evidence of chunk learning in continuous serial reaction-time tasks?". *Psychological Research*, 72, 2008, pp. 387–396.

Kieras, David / Meyer, David E.: »An overview of the EPIC architecture for cognition and performance with application to human-computer interaction.«. *Human Computer Interaction*, 12, 1997, pp. 391–438.

Krivec, Jana / Guid, Matej / Bratko, Ivan: "Identification and characteristic descriptions of procedural chunks.". In Dini, Petre (Ed.). *Proceedings of Computation world 2009, Computation world: future computing, service computation, adaptive, content, cognitive, patterns.* New York: IEEE Coputer Society, 2009, pp. 448–453.

Langley, Pat / Choi, Dongkyu: »Learning recursive control programs from problem solving.«. *Journal of Machine Learning Research*, 7, 2006, pp. 493–518.

Logan, Gordon D.: "Toward an Instant Theory of Automatization.". *Pyschological Review*, 95, 1988, pp. 492–527.

Lu, Chun-Hung / Wu, Chia-Wei / Wu, Shih-Hung / Chiou, Guey-Fa / Hsu, Wen-Lian: "Ontological Support in Modeling Learners' Problem Solving Process.". *Educational Technology & Society*, 8(4), 2005, pp. 64–74.

MacGregor, James N.: »*Short-term memory capacity: Limitation or optimization?*«. *Psychological Review*, 94 (1), 1987, pp. 107–108.

McFarlane, Angela / Sparrowhawk, Anne / Heald, Ysanne: *Report on the educational use of games: An exploration by TEEM of the contribution which games can make to the education process.* Cambridgeshire, UK 2002.

Miller, George A.: "The Magical Number Seven, Plus or Minus Two: Some Limits on our Capacity for Processing Information.". *Psychological Review*, 63, 1956, pp. 81–97.

Mitrovic, Antonija / Mayo, Michael / Suraweera, Pramuditha / Martin Brent: "Constraint-Based Tutors: A Success Story. ". In: Monostori, Laszlo / Váncza, József / Ali Moonis (eds.): *Proceedings of the 14th International conference on Industrial and engineering applications of artificial intelligence and expert systems: engineering of intelligent systems (IEA/AIE '01).* Springer-Verlag: London 2001, pp. 931–940.

Mitrovic, Antonija / Martin, Brent / Suraweera, Pramuditha / Zakharov, Konstantin / Milik, Nancy / Holland, Jay: "ASPIRE: An authoring system and deployment environment for constraint-based tutors. ". *International Journal of Artificial Intelligence in Education*, 19, 2009, pp. 155–188.

Nkambou, Roger / Mizoguchi, Riichiro / Bourdeau, Jacqueline: *Advances in intelligent tutoring systems.* Heidelberg: Springer 2010.

Ohlsson, Stellan: Learning from Performance Errors. *Psychological Review*, 103 (2), 1996, pp. 241–262.

Psotka, Joseph / Mutter, Sharon A.: *Intelligent Tutoring Systems: Lessons Learned.* Lawrence Erlbaum Associates 1988.

Ritter, Steven / Kulikowich, Joanna / Lei, Pui-Wa / McGuire, Christy / Morgan, Pat: "What evidence matters? A randomized field trial of Cognitive Tutor Algebra I.". In: Hirashima, Tsukasa / Hoppe Ulrich / Young Shelley Shu (eds.), *Supporting Learning Flow through Integrative Technologies*, 162. Amsterdam: IOS Press 2003, pp. 13–20.

Self, John: »The defining characteristics of intelligent tutoring systems research: ITSs care, precisely.«. *International Journal of Artificial Intelligence in Education*, 10, 1999, pp. 350–364.

Sowa, John F.: *Knowledge Representation: Logical, Philosophical, and Computational Foundations.* Pacific Grove, CA, USA: Brooks Cole Publishing 1999.

Simon, Herbert / Chase, William: "Perception in chess.". *Cognitive Psychology*, 4, 1973, pp. 55–81.

Vanlehn, Kurt / Lynch, Collin / Schulze, Kay / Shapiro, Joel A. / Shelby, Robert / Taylor, Linwood/ Treacy, Don / Weinstein, Anders / Wintersgill Mary: "The Andes Physics Tutoring System: Lessons Learned". Int. J. Artif. Intell. 15(3), 2005, pp. 147–204.

Wickelgren, Wayne A.: »*Single-trace fragility theory of memory dynamics.*«. *Memory & Cognition*, 2, 1974, pp. 775–780.

Wyer, Robert S. Jr. / Srull, Thomas: *Handbook of social cognition: Vol. 1, Basic processes (2nd ed.)*. Hilsdale, NJ: Erlbaum 1994, pp. 1–40.

Jernej Agrež
Faculty of Information Studies in Novo mesto
Ljubljanska cesta 31a, p. p. 603, 8000 Novo mesto, Slovenia
jernej.agrez@fis.unm.si

Nadja Damij
Newcastle Business School, Northumbria University
Faculty of Business and Law
Newcastle upon Tyne, NE2 1XA, England
nadja.damij@northumbria.ac.uk

Knowledge Dynamics in a State of Emergency: A Case Study

Abstract: *In this paper, we set our focus on the information society on the community level in connection with information, learning and knowledge exchange. For purposes of this research, we used several different methods, binded in the information technology toolkit. We carefully studied regional and local standard operating procedures in case of flood events. To be able to continue with the research we used statistical methods to compare four flood events in the region. We collected information about entities in distress from the Administration for civil protection and disaster relief's database and created a fuzzy system for assessing learning and information sharing capacity of the communities. Further, we conducted interviews with the members of households (entities), who directly faced at least two flood events and the vice commander of local civil protection, to be able to assess entities' learning and assess the presence of microscale information society. Finally, we used new insights to design learning based social information communities, incentives, based on business process management, using TAD methodology and the iGrafx simulation tool. The paper provides us with the insight of, how important it is to have a stable microscale information society network, present in the scope of local flood response activities. At the same time, we revealed the link between the learning and microscale information society. Not only is experiential learning important for fostering information communities during emergencies, but also a more systematic approach to flood resilience learning and self-reliance is important.*

Keywords: *knowledge management, learning, emergency, floods*

1. Introduction

The information society has aims, objectives, consistent structure, and possesses an openness (Ziemba 2013, pp. 106–117). Trying to identify its place among the ongoing social trends, rapidly developing technology, course of cultural changes and actions of policy makers, we find the information society integrated in all of the listed fields. Even more, the information society became an inevitable part of national strategic guidelines in developed societies (Elyakov 2010, pp. 60–73). Therefore, we find the organizations, which act in the compliance with the lawfulness of the information society, different from those, which are ignoring it. Differences do not occur only when compared with their vision and mission, but also in their legal form and working methods.

Legislation in Slovenia partially divides information-oriented from the no information-oriented sphere. On the one hand government agencies and some entities of the public sector cannot avoid using information technology as the tool to provide their services to the citizens. The business sector is obliged to use such information technology based services, while citizens remain encouraged to do so if they have the knowledge and the wish. On the other hand individual in ones' daily life uses information by his own choice, not by government regulation, consequently building the information society more or less knowingly.

A state of emergency, which occurs because of a natural disaster or other types of incident with human or material casualties, triggers the emergence of informal entities such as local humanitarian and responding initiatives. Such organizational systems take an important role in the information society, supporting formal entities, or even freelancing their own action plan. This case study refers to the microscale information society phenomenon. We focused the research on four different flood events, which triggered a state of emergency to the local level in the Lower Sava Valley, in SE Slovenia.

There are several local communities in the Lower Sava Valley, which is a geographical and statistical region. In fact, the valley is a lowland caught between two hill ridges on the southern and eastern side. We set a single local community to be an entity, composed of several micro entities (flood-endangered households). We assessed knowledge transfer on the micro scale, by measuring learning capacity of every micro entity. Further, we assessed knowledge transfer from every micro entity towards the surrounding institutional response system and from the surrounding institutional response system towards the micro entities.

In the context of regional disaster management system, we detected an emergency response to demand from the flood-endangered communities. Major flood events between 2009 and 2015 revealed that floods rarely affected single households. Instead, most communities are affected and in need of a response. A Slovenian system for protection from natural and other disasters supports the communities in distress by means of emergency phone services and regional

Civil Protection Command, which control response activities at the regional level. Beside these local Civil Protection and fire brigades, take care of localized disaster events. The system is based on national strategic documents, as well as regional and local operational procedures that are based on experiences with natural disasters at the regional or local levels.

Community experiential learning represents an important possibility for the development of the microscale information society, to support strategic operative policies. Communities and entities learn through their own experience with flood risks. Entities share the information they collect about local water conditions and their own endangerment with the responsible public bodies, but even more important is their growing capacity of self-sustaining response to flood events. Such capacity, based on the micro scale social information society, brings the possibility to influence official flood response procedures.

2. Information Society, Information Communities and Information Technology in Connection with Emergency Knowledge Dynamics

Global climate change has increased the frequency of extreme precipitation events and may cause many more floods in the future. This brings great human life and property losses by flooding fields, washing away housing, infectious diseases, etc. (Qi et al 2014, pp. 3787–3802). Such change brings also an additional need for more efficient social reaction, achieved by the use of information technology. Flood risk is expected to increase in many regions of the world (Cammerer et al. 2013, pp. 1243–1270), which indicates heavier exposure to consequences caused by flooding. Repeated exposure to natural disasters, such as floods, renders people more resilient and facilitates social connectedness that enhances a sense of place (Boon 2014, pp. 683–701). A logical result of threat acceptance and rising resilience in the urban, suburban, and rural communities in flood-endangered areas is the development of risk management, capable of flexible adaptation and improvement (Zevenbergen et al. 2013, pp. 1217–1225), which holds connections with organizations, present in the field of information technology and society.

Even though the disaster risk management framework is under the jurisdiction of national and local governments from the financial (Chang, Berdiev 2013, pp. 406–448), legal (Nishikawa 2010, pp. 249–262), and implementation (Kakumoto et al. 2013, pp. 55–64) perspectives, suburban and rural communities tend to rely on their own experience and information sharing (Ryan 2013, pp. 229–242), creating their own informal field of microscopic information communities. Suburban and rural communities are more likely to assess an ongoing situation and build their response based on their own social network, which includes trustworthy information sources (Hagar 2010, pp. 38–44). In addition to such community information networking, suburban and rural communities show sensitivity to the geographical origin of the external information concerning the ongoing flood situation. Communities are more receptive to the external information from local origin than information from afar (Cohen 2007, pp. 88–96).

Microscale information communities during and after flood events enhance disaster preparedness not only of private households affected by floods, but businesses and authorities as well (Kreibich et al. 2011, pp. 59–71). Supported by experiential learning, the emergence of such communities as we see it is a knowledge supported process, where experience plays a central role (Kolb et al. 2000). For such processes, it is of great importance to be implemented in the community of interconnected social partnerships (Benecke, Bezuidenhout 2011, pp. 55–69) because community-based initiatives, which are experiential and action oriented, complement regular forms of learning (Jakubowski, Burman, 2004, pp. 160–177). They provide learners with the chance to participate in organized microscale networks and to meet the social needs of the community (Tapps 2014).

Experiential learning is a very effective approach to enhance the microscale information communities within the disaster threatened learning society (Rijumol et al. 2010). Well-managed and applied experiential community knowledge can play a vital role through ensuring the availability and accessibility of accurate and reliable disaster risk information when required (Pathirage et al. 2012, pp. 237–252) which consequently applies to the emergence of the microscale information community network. In general, people tend to ignore personal disaster risks, representing themselves as disaster immune (Slovic et al. 1981, pp. 17–34). Such optimism is present in communities that are threatened with low probability disaster risks and where people lack direct experiences (Weinstein 1989, pp. 1232–1233). Experiential learning thus improves risk perception in general and flood risk perception in particular (Bosschart et al. 2013, pp. 1661–1680), (Grothmann, Reusswig 2006, pp. 101–120), (Siegrist, Gutscher2006, pp. 971–979), (Terpstra, Lindell 2012, pp. 993–1018) and therefore encourages the emergence of a microscale information community.

Experiential community learning which occurs as single-loop learning results on the one hand in the changes of the community behavior, strategies, and techniques (Hayward et al. 2007, pp. 239–250). On the other hand, it enhances social, economic networking among communities and public services, which are responsible for disaster preparedness and response (Gamboa-Maldonato et al. 2007, pp. 239–250). Experiential learning enables a community to verify its expectations of the governing bodies' capabilities to implement an efficient disaster response process through its ability to influence the response process (Chamlee-Wright,

Storr 2010, pp. 253–274). Such mutual social, economic networking during disaster events is of high importance for a community in terms of developing its resilience and withstanding the burden of the ongoing emergency (Plough et al. 2013, pp. 1190–1197). For this reason, community reactions have a crucial role in microscale information communities` optimization, during disaster events (Shughart 2011, pp. 519–539).

Efficient inclusion of information communities, information society and information technology in the disaster risk management are a shared goal of the community, national and local governments, and disaster responding services (Metcalfe et al. 2012), (Reaves et al 2012, pp. 60–68). It is also a subject of active community participation in the disaster management cycle (Jahangiri et al. 2011, pp. 82–94). To achieve a solid and sustainable level of community-based safety, a community should encourage its individual members to participate in microscale information community networking and disaster management processes in order to develop knowledge based prevention capabilities and risk reduction (Xi et al. 2013, pp. 2215–2239). An active community has strong learning abilities. Therefore, it can easily assess its needs, share information, and provide necessary public influence (Petterson 2010, pp. 127–141). A learning community is capable of overtaking traditional organizational structures and mechanisms when addressing disaster and risk management (Said et al. 2011, pp. 266–280) through a diverse range of organizational and professional resources that can be called upon to assist recovery (Johnston et al. 2012, pp. 252–268).

3. Information Technology Toolkit for Emergency Knowledge Dynamics Assesment

The research methodology toolkit that made it possible to determine the community-learning performance consisted of three stages. First, we gathered data from the Administration for Civil Protection and Disaster Relief database and compared it with the data from the logbook of a local Civil Protection Command. The data covered four flood events from 2010 to 2014, providing official responses for 185 distress events including 167 entities. We merged the data in a single database, using each distress event of each entity as an instance in the database. Further, we obtained the data on river flow rates and rainfall from the Slovenian Environment Agency, and merged that with a previously created database, using the dates of the distress events as merging criterion. Finally, we used the Google Maps and Google Earth API to determine the altitude of each entity and the distance from the distress source. The attributes of the final database are presented in Table 1 and Table 2.

Table 1: Attributes of the final database 1

Date	Year	Distress type	Municipality	Community	Entity	Distance	Altitude
13.9.2014	2014	rainfall	KRŠKO	ARDRO PRI RAKI	ARDRO PRI RAKI 12	0	222.00
18.9.2010	2010	rainfall	BREŽICE	ARTIČE	ARTIČE 40	0	217.00
13.9.2014	2014	rainfall	BREŽICE	BREŽICE	BREZINA 4A	0	152.00
18.9.2010	2010	Krka	KRŠKO	BROD V PODBOČJU	BROD V PODBOČJU 21	58	145.00

Table 2: Attributes of the final database 2

Activity	Responders	Flow rate Jesenice	Flow rate Podobočje	Flow rate Širje	Flow rate Hrastnik	Rainfall Bizeljsko	Rainfall Sromlje	Rainfall Brege	Rainfall Smednik
Water pumping	Local fire brigade	2274.00	446.00	697.00	978.00	42.4	46.7	82.4	80.6
Water pumping	Local fire brigade	2048.08	135.54	859.74	1481.55	72.6	64.4	102	75.5
Water pumping	Local fire brigade	2274.00	446.00	697.00	978.00	42.4	46.7	82.4	80.6
Saving the animals	Local fire brigade	2048.08	135.54	859.74	1481.55	72.6	64.4	102	75.5

The second step included semi-structured interviews with people who were part of the household that are represented in our database as entities. We selected 56 participants from flood-endangered entities. Participant recruitment criteria were: they had to be actively involved in at least one flood event, they had to live in one of the households affected by the detected distress sources, and they had to have open access to all the information about the ongoing situation, along with their family members and other members of the household. The participants were young adults and adults (30 male, 26 female) with the average age of 42,13 and standard deviation of 13,78. The aim of the interviews was to detect how entities perceived their own learning about floods, circumstances, and response actions of responsible services. We used an interview framework in a matrix form presented in Table 3 and analyzed the results with a paired t-test.

Table 3: Interview framework

	Risk awareness	**Disaster risk information**	**Household response**	**Knowledge source**
Before the first event	x	x	x	x
During the first event	x	x	x	x
After the first event	x	x	x	x
Before the following event	x	x	x	x
During the following event	x	x	x	x
After the following event	x	x	x	x

To be able to conduct a knowledge perception measurement, we developed a scale to determine whether disaster risk awareness, improved through the flood events, what was the information source of the entities, how household responded to the flood, and what was the perceived knowledge from the first flood events.

In the third step, we first compared rainfall and river flow attributes recorded in the four flood events with a multivariate analysis of variance, so as to be able to find a similarity of the flood events through flow rate and rainfall measurements, using Wilk`s lambda as a probability distribution. We compared the data by year of occurrence, trying to discover which attribute (flow rate in the first case and rainfall quantity in the second) influenced the difference most significantly. Based on the events' similarity, and learning rules from the third step we designed a fuzzy system (Hornik, Meyer 2009, pp. 1–27) presented in Figure 1, which served as a tool for measuring the learning performance of flood-endangered communities.

Figure1: Fuzzy system

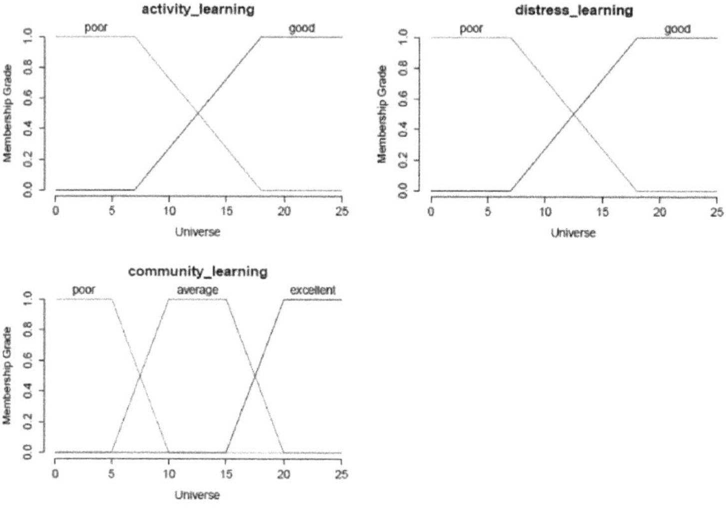

Finally, we designed an optimization algorithm that simulates additional systemic community learning, besides the experiential learning. We collected the optimization rule basis from the Resolution of national security strategy of the Republic of Slovenia (National Assembly, Republic of Slovenia 2010) and the Resolution of the national program of protection against natural and other disasters from the years 2009 to 2015 (National Assembly, Republic of Slovenia, 2009). The optimization algorithm foresees several educational prevention measures, to be able to increase resilience and self-reliance of the flood-endangered communities: Improvement of general flood preparedness; Systematic modernization of training programs; New educational programs for pre-school and school children; Encouragement of measures that rise prevention, protection and security awareness.

We applied simulation of educational prevention measures on the flood events` data, including following process dimensions: communication time, travel distance, number of process architectures, number of process patterns, number of activities, number of entities in distress, total number of executed standard operating procedures, and number of different standard operating procedures during one event. We took into account the severity level of every single assessed flood event. According to the recorded occurred damage, the scope of the flooding and the response force activity workload, we determined the event horizon when an individual household becomes incapable of providing self-reliant flood protection and must request for an emergency response.

4. Knowledge Dynamics in a State of Emergency Assesment

We compared hydrological and meteorological data of four floods (period of four days) through the multivariate analysis of variance and by the simple quantitative comparison of variables. Table 4 compares the Wilk`s Lambda values from flow rate MANOVA testing. The results of the comparison of flood events in the year 2012 vs. 2014 and 2013 vs. 2014 show significant difference only in the Podbočje station flow rate. Furthermore, a nearly significant difference was identified with the regard to the Hrastnik station flow rates in the year 2010 vs. 2013 and 2012 vs. 2013. There was no significant difference in the comparison between 2010 vs. 2012 and 2010 vs. 2014.

We also used rainfall data, analyzed with the multivariate analysis of variance and by quantitative comparison. MANOVA (Table 5) revealed no significant difference in the variance among the four day rainfall measurements during floods. Therefore, we determined similarities, based on quantitative comparison, which revealed similarities between the events in 2010 vs. 2014 and 2013 vs. 2014.

Table 4: MANOVA, river flow data

		Jesenice station, flow rate		Podbočje station, flow rate		Širje station, flow rate		Hrastnik station, flow rate	
Year	p	Wilk`s λ	*F-value*	Wilk`s λ	*F-value*	Wilk`s λ	*F-value*	Wilk`s λ	*F-value*
2010 2012	0.05	0.463	0.594	0.5322	0.4262	0.4612	0.5991	0.4392	0.6626
2010 2013		0.1406	26.741	0.6564	0.2135	0.1752	22.127	0.05837	48.703
2010 2014		0.4884	0.5275	0.3651	0.9218	0.4267	0.7012	0.14	2.684
2012 2013		0.2091	18.656	0.7397	0.1184	0.357	0.9553	0.07379	42.283
2012 2014		0.9823	5,00E-04	0.009884	11.312	0.9175	0.0114	0.2932	12.656
2013 2014		0.2491	15.448	0.03464	64.582	0.4425	0.6528	0.4221	0.7159

Table 5: MANOVA, rainfall data

		Bizeljsko		Sromlje		Brege		Smednik	
Year	**p**	**Wilk`s λ**	***F-value***	**Wilk`s λ**	***F-value***	**Wilk`s λ**	***F-value***	**Wilk`s λ**	***F-value***
2010 **2012**	0.05	0.2164	1.8009	0.1809	2.1483	0.2095	1.8623	0.1953	1.9973
2010 **2013**		0.3229	1.11	0.3925	0.8169	0.3217	1.1155	0.3583	0.9499
2010 **2014**		0.3715	0.8962	0.5047	0.4877	0.6111	0.28	0.6233	0.261
2012 **2013**		0.4814	0.5452	0.2351	1.6483	0.4438	0.6488	0.2283	1.7019
2012 **2014**		0.5559	0.3777	0.239	1.6184	0.3442	1.0108	0.3301	1.075
2013 **2014**		0.9379	0.0065	0.7901	0.0758	0.5736	0.3441	0.6519	0.2195

		Kostanjevica		Planina		Cerklje ob Krki	
year	**p**	**Wilk`s λ**	***F-value***	**Wilk`s λ**	***F-value***	**Wilk`s λ**	***F-value***
2010 **2012**	0.05	0.2242	1.7355	0.2576	1.4859	0.2397	1.6131
2010 **2013**		0.5038	0.49	0.637	0.2406	0.4857	0.5342
2010 **2014**		0.6476	0.2255	0.7688	0.0925	0.8246	0.0525
2012 **2013**		0.2728	1.3869	0.2708	1.3994	0.3252	1.0987
2012 **2014**		0.3538	0.9687	0.337	1.0429	0.378	0.8711
2013 **2014**		0.8443	0.0412	0.8694	0.0288	0.6869	0.1747

We integrated the results of the flood similarity, into the fuzzy system, implemented in the R programming environment. To be able to create an evaluation data set, we gathered the emergency response data from the database, governed by the Slovene Administration for civil protection and disaster relief. We extracted the data, which included flood response in communities, during the floods, previously detected as similar.

We divided the data into community sets. The structure of every set was the following: S = {A, B, C, D, E}.

- Subset A represents the distress source,
- Subset B represents the number of distress cases during the first flood,
- subset C represents the number of distress cases during the second flood,
- subset D represents the primary response activities during both flood events,
- subset E represents the secondary response activities during second flood event.

The following stage of the evaluation included the categorical evaluation in which the learning performance of the communities was measured using the condition of the same distress source during both flood events, and the following five criteria:

- $x1 \in B \cdot x2 \in C \Rightarrow L1 = \{1\}$, distress is detected during both flood events;
- $x1 \in B \Rightarrow L2 = \{1\}$, distress is detected only during the first flood event;
- $x2 \in C \Rightarrow L3 = \{1\}$, distress is detected only during the second flood event;
- $x3 \in D` \cdot x3 \in D`` \Rightarrow L4 = \{1\}$, primary response activities are the same during both flood events
- $x4 \in E` \cdot x4 \in E`` \Rightarrow L5 = \{1\}$, secondary response activities are the same during both flood events.

The subset $X = \{x1, x2, x3, x4\}$ represents data typical for a single entity (communities are built out of entities, which are in reality flood endangered households), while the subset $L = \{L1, L2, L3, L4, L5\}$ represents its learning grades. To be able to place the grades of the entities into the community perspective, we summed them into community subsets $CS_n = \{\sum_{L1n}^{n} L1,\ \sum_{L2n}^{n} L2,\ \sum_{L3n}^{n} L3,\ \sum_{L4n}^{n} L4,\ \sum_{L5n}^{n} L5\}$ $\Rightarrow CS_n=\{CL1n, CL2n, CL3n, CL4n, CL5n\}$ and further weighted them with the weighting rules presented in Table 6.

Table 6: Weighting criteria

	CL1n		CL2n		CL3n		CL4n		CL5n	
Value ranks	**Value**	**Weight**	**Value**	**Weight**	**Value**	**Weight**	**Value**	**Weight**	**Value**	**Weight**
High rank	3	8	<=16	8	<=16	8	>10	15	>1	15
Middle rank	2	4	< 6	4	< 6	4	10	10	1	10
Low rank	1	1	1	1	1	1	<10	10	-	-

The community learning analysis provided us with the insight, how poor is the overall learning process of communities about floods in order to protect themselves and gain higher resilience. Out of 59 communities, only 10 were graded with the grade higher than poor. Most communities, even though they faced two similar floods in a period of four years, have not used the first experience to learn about how and when the flood would threaten them and how to protect their property against rising water. Those communities, which demonstrated learning process, graded between good and excellent, gain their knowledge based only on the experiential learning. No other, more formal knowledge source or learning process that would provide them with knowledge that is more explicit, took place. Semi-structured interviews with the endangered entities provided us with the nominal scale, presented in the Table 7.

Table 7: Interview data nominal scale

	Detected possibilities				
Field of inquiry	1	2	3	4	5
Disaster risk awareness	No	Present	Moderate	Influential	High
Disaster risk information source	Media	Observation	Government agencies	Public services	/
Household response	Observation	Preparedness	Response	Recovery	Risk reduction
Knowledge source	Past experiences	Ongoing experience	/	/	/

We used the scale to represent different aspects of the flood event perception before, during and after the two floods, which affected entities that participated in the research.

General overview of the mean values, calculated from the nominal data, retrieved from the 56 interviews, using the framework is presented in Table 7. We gained additional understanding of the interview data, using paired t-test, presented in Table 8. The results of the test revealed that the significant differences discovered were disaster risk awareness before and during the first and the second flood events. Also the difference of the information source during first and second flood events is shown as significant. Other fields of inquiry, including knowledge source, were not significantly different.

An analysis of the interview data, including t-test, revealed that during and after the floods, the entities in the flood-affected communities used mostly real-time experiential learning, based on the ongoing situation. Before the floods, entities` knowledge source was mostly past experiential learning.

Table 8: Mean, standard deviation values and paired t-test of interview data

Field of inquiry			First flood event	Second flood event	Paired t-test: t	p	Interpretation
Before	Disaster risk awareness	Mean	2,05	2,45	-3.578	0.001	Significant
		SD	0,77	0,58			
	Disaster risk information	Mean	1,68	1,95	-2.027	0.224	Not significant
		SD	0,47	0,82			
	Household response	Mean	1,36	1,41	-1.000	0.20	Not significant
		SD	0,48	0,49			
	Knowledge source	Mean	1,32	1,05	2.806	0.55	Not significant
		SD	0,47	0,21			
During	Disaster risk awareness	Mean	3,27	3,50	-1.418	0.001	Significant
		SD	1,09	0,94			
	Disaster risk information	Mean	2,00	2,41	-2.001	0,001	Significant
		SD	0,52	0,72			
	Household response	Mean	2,41	2,41	0	0,31	Not significant
		SD	0,72	0,72			
	Knowledge source	Mean	1,91	2,00	-1.559	0.199	Not significant
		SD	0,29	0,00			
After	Disaster risk awareness	Mean	2,68	2,68	0	0,24	Not significant
		SD	0,92	1,06			
	Disaster risk information	Mean	1,09	1,14	-0.568	0.470	Not significant
		SD	0,29	0,46			
	Household response	Mean	1,82	1,91	-1	0.15	Not significant
		SD	1,34	1,62			
	Knowledge source	Mean	1,91	1,82	1	0.598	Not significant
		SD	0,29	0,49			

Flood event response process model, is built of 23 activities and eight entities. The model includes all key decision points, which influence the course of the activity flow and knowledge exchange as well. With conversion into a discrete simulation and execution of 33 predefined scenarios, we were able to collect activity transaction data, linked with not only the key entities, but also to their decision making. To be able to give a general assassment of the transaction dynamics, we calculated mean values and standard deviations of the transaction count, where we included the key entities of the flood event response process and their decision making activities (Table 9).

Table 9: Transaction count, mean values and standard deviation

	mean	SD
community member transactions	8,85	0,43
community member positive decisions	3,61	0,89
community member negative decisions	1,30	0,67
distress call center transactions	5,00	2,42
distress call center positive decisions	0,36	0,48
distress call center negative decisions	0,18	0,39
CP command transactions	7,42	2,19
CP command positive decisions	1,33	1,03
CP command negative decisions	0,33	0,47
CP response transactions	0,39	0,78
fire brigade command transactions	6,58	2,09
fire brigade command positive decisions	1,09	1,00
fire brigade command negative decisions	0,48	0,50
fire brigade transactions	0,70	1,09
road company transactions	0,18	0,57
geotechnician transactions	0,55	0,89
number of red knowledge links	2,27	1,64
number of blue knowledge links	2,42	0,92
number of knowledge links	4,70	2,39

We present a comparison between proportions of mean values and standard deviation values in Figure 2. Comparison provides us with an interesting insight, how the entities and their decision differ through the 33 simulated scenarios. Both, Table 9 and Figure 2 show us the highest average number of community member transactions with a small standard deviation. We can find the closest ratio, only with a smaller transaction count average value, and a higher standard deveiation value, by the community member positive decision transaction count, followed by the community member negative decision transaction count. A similar pattern follows all key entities with their decision making transactions.

Figure 2: comparison between proportions of mean values and standard deviation values

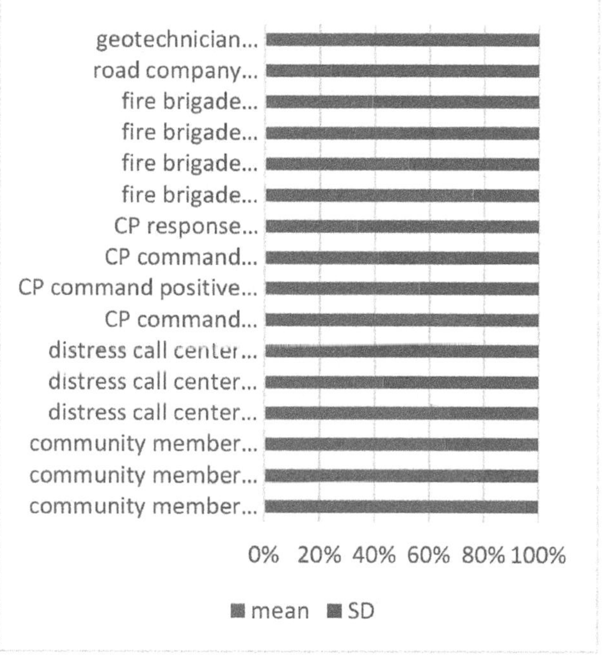

Figure 3: Comparison of knowledge exchange transactions

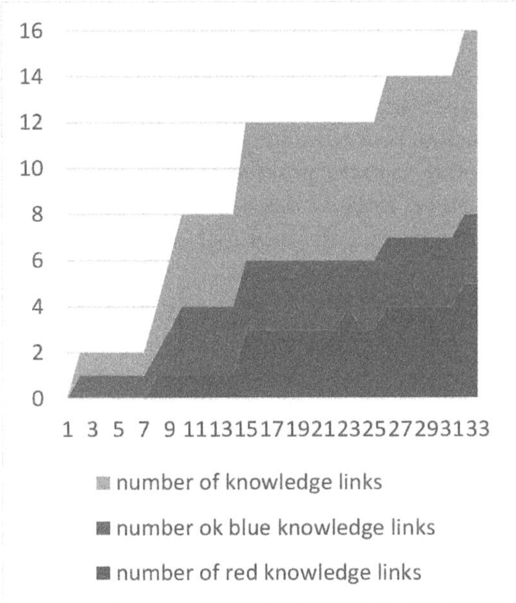

Beside the transaction count focused on entities and their decision making, we conducted also transactions, which directly indicate knowledge exchange among two groups of entities: customers (flood endangered entities) and providers (flood responding organizations), presented in Figure 3. We named red knowledge links the transactions which indicate knowledge exchange on relation customers to provider. Blue knowledge links are those transactions, which indicate knowledge exchange on relation provider to customer. Within the group of 33 simulated scenarios we were able to detect a minor group of six scenarios, where knowledge exchange runs only in relation of provider to customer. The other 27 scenarios demonstrate knowledge exchange in both relations. Graphical representation also indicates several groups of scenarios which show similar characteristics, refering to knowledge exchange.

5. Conclusion

The microscale information society holds an important role in the scope of self-relying flood protection. To be able to establish social, economic networking, communities face the inevitable requirement of reliable knowledge flow. At the same time it is of high importance, to provide communities with the opportunity to learn not only from the past experiences, but also in a form of systematic approach to learn systematically about flood endangerment prevention procedures, flood emergency measures, use of equipment, communication and other necessary measures. The microscale information society has the potential to take over a higher share of institutional flood response, which could eventually relieve the official disaster response organization and therefore optimize usage of public resources, including human resources, equipment, consumers and work time.

Self-relying flood protection is based on the knowledge flow among entities who form the microscale information society network on the one hand, and on the other hand it relies on the knowledge flow between the endangered communities and surroundings, composed out of surrounding disaster relief organizational system. The potential, held by the microscale information society network is shown to be much higher than the applied information sharing measures. The main factor, influencing the low efficiency is the low dynamics of the knowledge transition between flood endangered communities and support surroundings. We can see that the learning process between the two sides is not fully mutual. Different flood reaction scenarios bring different learning processes and consequently different output knowledge.

In the scope of our research, knowledge is the main attribute which influences and encourages the miscroscale information society. Even with a low knowledge level detected, flood endangered communities are capable of establishing an information society network to some extent. The information society links individual entities into the community net, which is capable of more coordinated and efficient action, than single entities. Even though single entities posses experiential knowledge, it is of high importance for entities to connect in order to create a knowledge platform for the growth of the microscale information society. An interesting phenomena which relates to the microscale information society in this case is that it does not exceed the duration of the emergency situation with short term mitigation activities. Consequently, it emerges in waves, depending on the level of the emergency. To be able to maintain a long term information communitiy networking platform, it would be necessary to maintain knowledge flow among entities, which form communities and between communities and surrounding flood support system.

A limitation of our research is linked to the specificity of the emergency, which triggers the microscale social economic emergence. It is unclear and impossible to foretell with certainty, whether research results could be applied to other emergency situations within the scope of a natural disaster. Compared to natural disasters such as an earthquake which in its most devastating form occurs rarely but with high destruction, flood events occur frequently but with less devastating effects than an earthquake, especially when we compare massive earthquakes like Peru, Haiti or Sumatra to the flood events in Lower Sava region. Known limitations lead us to the research question, whether it is possible to compare the microscale information society which emerges during different emergency situations and what are the differences between the detected microscale information society phenomena.

6. References

Benecke, D., Rene / Bezuidenhout, Rose-Marie: "Experiential learning in public relations education in South Africa". *Journal of Communication Management,* 15(1), 2011, pp. 55–69.

Boon, J. Helen: "Disaster resilience in a flood-impacted rural Australian town". *Natural Hazards,* 71(1), 2014, pp. 683–701.

Bosschart, Adwin / Kuiper, Wilmar / van der Schee, Jup / Schoonenboom, Judith: "The role of knowledge in students' flood-risk perception". *Natural Hazards,* 69(1), 2013, pp. 1661–1680.

Cammerer, Holger / Thieken, H., Annegret / Verburg, Peter: "Spatio-temporal dynamics in the flood exposure due to land use changes in the Alpine Lech Valley in Tyrol (Austria)". *Natural Hazards,* 68(1), 2013, pp. 1243–1270.

Chamlee-Wright, Emily / Storr, V. Henry: "Expectations of government's response to disaster". *Public Choice,* 144(1), 2010, pp. 253–274.

Chang, P. Chun / Berdiev, N. Aziz: "Natural Disasters, Political Risk and Insurance Market Development". *The Geneva Papers,* 38(1), 2013, pp. 406–448.

Cohen, Erez / Hughes, Peter / White, B. Peter: "Media and bushfires: A community perspective of the media during the Grampians Fires 2006". *Environmental Hazards,* 7(1), 2007, pp. 88–96.

Elyakov, A. D.: "The Nature of the Modern Information Society". *Scientific and Technical Information Processing,* 37(1), 2010, pp. 60–73.

Gamboa-Maldonato, Thelma / Marshak, H., H. / Sinclair, R. / Montgomery, S. / Dyjack, D.: "Building Capacity for Community Disaster Preparedness: A Call for Collaboration Between Public Environmental Health and Emergency Preparedness and Response Programs". *Environmental Health,* 75(2), 2012, pp. 24–29.

Grothmann, Torsten / Reusswig, Fritz: "People at risk of flooding: Why some residents take precautionary action while others do not". *Natural Hazards,* 38(1), 2006, pp. 101–120.

Hagar, Christine: "Farmers' search for information during the UK foot-and-mouth disease crisis- what can we learn?". *Australian Journal of Emergency Management,* 25(4), 2010 pp. 38–44.

Hayward, Graeme / Diduck, Alan / Mitchell, Bruce: "Social Learning Outcomes in the Red River Floodway Environmental Assessment". *Environmental Practice,* 9(4), 2007, pp. 239–250.

Hornik, Kurt / Meyer, David: "Generalized and customizable sets in R". *Journal of Statistical Software,* 31(2), 2009, pp. 1–27.

Jahangiri, Katayoun / Izadkhah, O. Yasamin / Tabibi, Jamaledin: "A comparative study on community-based disaster management in selected countries and designing a model for Iran". *Disaster Prevention and Management,* 20(1), 2011, pp. 82–94.

Jakubowski, Lisa Marie / Burman, Patrick: "Teaching community development: a case study in community-based learning". *Teaching Sociology* 32(2), 2004, pp. 160–177.

Johnston, David / Becker, Julia / Paton, Douglas: "Multi-agency community engagement during disaster recovery". *Disaster Prevention and Management,* 21(2), 2012, pp. 252–268.

Kakumoto, Shigeru, Furoto, T. / Yamada, H. / Sasaki, M. / Shiwaku, K.: "Proposal to Realize Rarmis (Risk Adaptive Regional Management Information System) Concept by Spatial Temporal Information System Dimsis-ex and Some Case Studies — Toward Collaborative Realization of Common Software (SYSTEM) Platform for Disaster Prevention (CSPDP)". *Asian Journal of Environment and Disaster Management,* 3(1), 2013, pp. 55–64.

Kolb, A. David / Boyatzis, Richard / Mainemelis, Charlampos: Experiential Learning Theory: Previous Research and New Directions, Perspectives on cognitive, learning, and thinking styles, Lawrence Erlbaum, New Jersey, 2000.

Kreibich, Heidi / Seifert, I. / Thieken, A. H. / Lindquist, E. / Wagner, K./ Merz, B.: "Recent changes in flood preparedness of private households and businesses in Germany". *Regional Environmental Change,* 11(1) 2011, pp. 59–71.

Metcalfe, Victoria / Haysom, Simone / Gordon, Stuart: "Trends and challenges in humanitarian civil–military coordination: a review of the literature". The Humanitarian Policy Group: London, 2012.

National Assembly, Republic of Slovenia, Resolution of national security strategy of Republic of Slovenia, *Uradni List RS,* Vol.92, No.7, 2010.

National Assembly, Republic of Slovenia, Resolution of national program of protection against natural and other disasters from year 2009 to year 2015, *Uradni List RS,* Vol.51, No.6, 2009.

Nishikawa, Sastoru: "From Yokohama Strategy to Hyogo Framework: Sharing the Japanese Experience of Disaster Risk Management". *Asian Journal of Environment and Disaster Management,* 2(3), 2010, pp. 249–262.

Pathirage, Chaminda / Seneviratne, Krisanti / Amaratunga, Dilanthi / Haigh, Richard: "Managing disaster knowledge: identification of knowledge factors and challenges". *International Journal of Disaster Resilience in the Built Environment,* 3(3), 2012, pp. 237–252.

Petterson, Olivia / Weil, Fredercik / Patel, Kavita: "The Role of Community in Disaster Response: Conceptual Models". *Population Research and Policy Review,* 29(1), 2010, pp. 127–141.

Plough, Alonzo / Fielding, J. E. / Chandra, A. / Williams, M. / Eisenman, D. / Wells, K. B. / Law, G. Y. / Fogleman, S. / Magana, A.: "Building Community Disaster Resilience: Perspectives From a Large Urban County Department of Public Health". *American Journal of Public Health,* 103(7), 2013, pp. 1190–1197.

Qi, Zhang / Jiquan, Z. / Liupeng, J. / Xingpeng, L. / Zhijun, T.: "Flood Disaster Risk Assessment of Rural Housings — A Case Study of Kouqian Town in China". *International Journal of Environmental Research and Public Health,* 11(1), 2014, pp. 3787–3802.

Reaves, J. Erik / Termini, Michael / Burkle, M. Frederick: "Reshaping US Navy Pacific Response in Mitigating Disaster Risk in South Pacific Island Nations: Adopting Community-Based Disaster Cycle Management". *Prehospital and Disaster Medicine,* 29(1), 2012, pp. 60–68.

Rijumol, K., C. / Thangarajathi, S. / Ananthasayanam, R.: "Disaster management through experiential learning". *Journal on Educational Psychology,* 4(1), 2010.

Ryan, Barbara: "Information seeking in a flood". *Disaster Prevention and Management,* 22(3), 2013, pp. 229–242.

Said, A. Mat, Ahmadun, F. / Mahmud, A. Rodzi / Abas, F.: "Community preparedness for tsunami disaster: a case study". *Disaster Prevention and Management,* 20(3), 2011, pp. 266–280.

Shughart, F: William: "Disaster Relief as Bad Public Policy". *The Independent Review,* 15(4), 2011, pp. 519–539.

Siegrist, Michael / Gutscher, Heinz: "Flooding Risks: A Comparison of Lay People's Perceptions and Expert's Assessments in Switzerland". *Risk Analysis,* 26(4), 2006, pp. 971–979.

Slovic, Paul / Fischhoff, B. / Lichtenstein, S.: "Perceived risk: psychological factors and social implications". *Proceedings of the Royal Society of London,* 1981, pp. 17–34.

Tapps, Tyler / Passmore, T. / Lindenmeier, D. / Kensinger, W.: "High School Physical Education Students and Experiential Learning in the Community:

A Classroom Assignment". *A Journal for Physical and Sport Educators,* 27(1), 2014, pp. 9–12.

Terpstra, Teun / Lindell, K. Michael: "Citizens' Perceptions of Flood Hazard Adjustments: An Application of the Protective Action Decision Model". *Environment and Behavior,* 45(8), 2012, pp. 993–1018.

Weinstein, D. Neil: "Optimistic Biases about Personal Risks". *Science, New Series,* 246(4935), 1989, pp. 1232–1233.

Xi, Zhang / Lixin, Yi / Dong, Zhao: "Community-based disaster management: a review of progress in China". *Natural Hazards,* 65(1), 2013, pp. 2215–2239.

Ziemba, Ewa: "The holistic and systems approach to the sustainable information society". *The Journal of Computer Information Systems,* 45(1), 2013, pp. 106–117.

Zevenbergen, Chris / van Herk, S. / Rijke, J. / Kabat, P. / Bloemen, P. / Ashley, R. / Speers, A. / Gersonius, B./ Verbeek, W.: "Taming global flood disasters. Lessons learned from Dutch experience". *Natural Hazards,* 65(1) 2013, pp. 1217–1225.

Dejan Valentinčič
School of Advanced Social Studies in Nova Gorica
Gregorčičeva 19, 5000 Nova Gorica
dejan.valentincic@fuds.si

Facebook as an Indicator of Vitality of Slovenian Minority and Diaspora Communities (with Emphasis on Young Population)

Abstract: *In this work we are examining the impact of the social network Facebook on Slovenian minority and diaspora communities. We are finding out if we can talk about virtual communities among these non-dominant ethnic groups and if Facebook is just a reflection or also impacting community lives of these groups and how does it impact ethnic identities. In the theoretical part we are presenting the role of social media (with an emphasis on Facebook) in modern information societies and summarising a short theory of cyber space and virtual communities. In the second theoretical chapter we are presenting the historical and present situation of Slovenian minority and diaspora communities. In the empirical part we are taking into consideration Facebook profiles of ninety young individuals from nine countries where most Slovenians live out of Slovenia. We are researching the language they use on Facebook, their activities and relations to local Slovenian communities and their interest for Slovenia. Interpreting these data we are finding out if young Slovenians from minority and diaspora communities use Facebook also as a tool for expressing their ethnic identity and if Facebook is just a reflection of realities or also a creator of new realities in these communities.*

Keywords: *minorities, diasporas, virtual communities, Facebook, personal profiles.*

1. Introduction

Ethnical diversity is a fact common to all modern societies. On the one hand there is a traditional complexity of state borders not matching with ethnic ones, on the other hand migrations make our societies more and more diverse. Even though the meaning of ethnic identifications is changing over time, it can not be denied that this is still an important factor which also causes conflicts. All democratic countries are (at least on the surface) obliged to respect the ethnic diversity of its population.

One fifth of people who have a Slovenian ethnic background lives outside of Slovenia. There is approximately 200.000 Slovenians living as traditional minorities in the neighbouring four countries (Austria, Croatia, Hungary and Italy) and approximately 300.000 people living as Slovenian emigrants and their descendants around the world. We can claim that all of those Slovenian communities try to perserve their ethnic identity, but in very different ways. This is the consequence of very different historical and present situations – the time of separation from the mother nation, the relation of the majority society, possibilities of relations with the mother nation, how large each ethnic community was, how many intellectuals it had, what was the level of rights that they were entitled to, etc. Since these historical conditions still determine today's situations of Slovenian communities abroad we present the historical development and today's organization of Slovenian communities in the third part of the article, before the empirical study.

In this article we are trying to identify and measure the vitality of the Slovenian minority and diaspora communities as it can be seen on the social network Facebook. Facebook is currently the most popular and widespread social network, which is also a tool of empowerment for its users. In the case of a non-dominant ethnic population the empowerment is often seen regarding the explicit expresions of an ethnic identity. Because of this Facebook is also an open book for researching the actual feelings and standpoints of members of specific ethnic communities. In the second part of the article we present what the theoretics say is a virtual community, before we try to find out in the empirical part if we can talk about virtual communities in cases of Slovenians abroad.

In the empirical part of the work we have chosen 10 young people from nine different countries: Austria, Croatia, Hungary and Italy as four countries with an indigenous Slovenian population and Argentina, Australia, Canada, Germany and USA as five countries with large Slovenian diaspora communities and examined their Facebook profiles. In all cases we took into consideration the period between 1st of October and 31st of December 2015.

All the examined profiles were chosen on the principle of randomness. We focused on the population in the age between 18 and 30. The reason for this is that the young population is the most active on social networks and at the same time they are the most important for the future of those communities. In our research we have deliberately left out individuals who have left Slovenia (first generation), especially in the last years, since their situations are completely different and usually they do not become active in Slovenian communities.

It is part of general knowledge that among Slovenian minorities in the neighbouring countries Slovenians in Italy are the most vital and most national awarned, Slovenians in Austria are in second place, where as those in Hungary and Croatia are less active. Among Slovenian disaspora communies Slovenians in Argentina are the most nationaly awarned, those in Germany should be in second place, where as those in anglo-saxon countries are more assimilated (those in Canada less than those in USA and Australia). In the empirical part we are testing if this is seen on Facebook too.

We are trying to answer the following research questions in the article: Does Facebook create new virtual communities among Slovenian diaspora and minority communities, or does it just reflect the reality? What is Facebook's reflection of vitality of the chosen Slovenian minority and diaspora communities?

2. Internet – Just an Indicator or also a Creator of Virtual Ethnic Communities?

In the last years the internet has become the central media, since it is the global communication infrastructure and at the same time the most spread forum, where discussions and negotiations on most different topics take place (Medvešek 2012, p. 9). The internet gives different possibilities of communication: one-to-one, one-to-many and many-to-many (Ibid.). Modern information and communication technologies made connecting people a lot easier than before. In the last years social networks have attracted more and more people, expecially the young ones. They allow users to present themselves and communicate in cyberspace with practically no additional tehnical or software skills (Boykova 2015, p. 1). Cyber space theoretics speak about »virtual communities«. If we study the topic from the ethnic perspective the situation is not any different at all. Here we can divide two different types of appcrances that are relevant for us: individuals who are connecting based on their ethnic background and individuals explicitly expressing their ethnic identifications in their profiles. Highlighting one's ethnic identity is usually connected with personal empowerment. Usually a population this feels marginalized has higher motivation to do that (Newsom & Lengel 2012, p. 33). In the cases of minority and diaspora communities the feeling of marginalization is not neccessary dependant on discrimination. The fact that their culture, language and identity is non dominant might be sufficient to perceive it as endangered.

Rheingold (1991, p. 5) defines virtual communities as a social aggregate which develops when a sufficient number of people continue with their disscusions long enough and with proper human feeling, with the purpose to create a network of human relations in the cyber space. The internet is not just the media of transport of information, but it can also be the symbol of development and the future, where also the minority language has its position (Medvešek 2009, p. 12).

Elkinns (1997, p. 141) claims that virtual ethnic communities are not any less real than usual ethnic communities, since they have the potential to become as significant for the identity of some people as existing ethnic communities are.

Medvešek (2009, p. 12) thinks that the internet can help with creating an ethnic identity, since it is easier for an individual or a community to keep connections with the population with common interests. Warschauer and De Florio-Hansen (2003, p. 22) think that it seems like the internet, in comparison with other mass media, does not give the opportunity to create a new modus of ethnic identities, it can only help with the increasing dynamics in the field of developing ethnic identifications.

A virtual community is free from some traditional criteria for the definition of at community: time and space (Čikić 2002, p. 85). Cohen (1985 in Jones 1999, p. 210) says that a virtual community is a community without physical limitations. Elkins (1997, p. 148) believes that virtual ethnic communities give the existing ethnic communities the oportunity to strenghten and refresh the community and help it to maintain its identity and activities, which would be otherwise difficult to keep through generations in small enclaves in the middle of other societies. Virtual ethnic communities give the individuals an opportunity for new ways of connecting between themselves (Čikić 2002, p. 85). The internet increases the possibilitiy of intensive communication waves, which is important for the future of an ethnic community (Medvešek 2009, p. 12).

Some claim that there is not just one world, but that simultaneously with the real life there is life in the cyber space, which supports maintaining old identities and, at the same time, creating new ones. It also gives the opportunity to maintain their ethnic identity for those, who live away from their ethnic community (Čikić 2002, p. 81). Verhlust (1999, p. 30) claims that technological innovations have, in many cases, brought the development of new diasporic identities, as the transformation of the old ones. Karim (1998 in Verhlust 1999, p. 31) says that we can speak about ethnic identity which is a common identity that is rising and is not an old identity, but a hybrid of the past connections, reestablished relations and experiences in the real life.

In today's cyber space social networks play the most imporant role for human relations, communication, spending free time and connecting with people with the same interest. They give people many opportunities for self-expression and communication through their online profiles. Social networks are today an integral part of the everyday life of millions of people (Boykova 2015, p. 1). The social network site Facebook has been explored as a tool for self-presentation of identity (Zhao et al 2008). As Facebook plays an imporatant role in the lives of individuals and different social groups, it is with no doubt the same also for ethnic communities. Besides being a tool for maintaining or even discovering its ethnic identity it can also give the communities new dynamics in their lives and activities.

The relation between virtual (on-line) and »real« (off-line) life are certainly complex and twisted. (Medvešek 2012, p. 13). If we focus just on Facebook as the most popular social network, three important questions come up:

- Does Facebook change the identities of members of the Slovenian minority and diaspora communities and communities themselves?
- Does Facebook (on personal profiles and in groups) just show the situation of Slovenian minority and diaspora communities as it is, or does it also affect the situation itself?
- Does Facebook have the same affect on all Slovenian minority and diaspora communities?

3. Short History and Actual State of Slovenian Minorities in the Neighbouring Countries

Indigenous Slovenian national minorities live in all four neighbouring countries (Italy, Austria, Hungary and Croatia). They were separated from their mother nation in different historical periods and the conditions for their development throughout history were different as well. Consequently, their life today varies greatly between individual countries and within different areas of individual countries. The manner in which the minorities have organized themselves differs completely as well. The first regions to be administratively separated from the mother nation were the valleys of Natisone, Torre and Resia, which were annexed to Italy in 1866 (Kacin Wohinz and Pirjevec 2000, p. 17). A year later, the Habsburg Monarchy reorganized itself into the Dual Monarchy of Austria-Hungary. The majority of Slovenian territory was placed under Austrian authority, except for Prekmurje and the Raba region, which were placed under Hungarian authority. Since then, the Raba region has never again been administratively united with the Slovenian ethnic territory (Zwitter 1990, pp. 484–488). After the First World War, Gaital in South Carinthia and the region of South Styria, where Slovenians live, were automatically placed under Austrian authority. It was left to the plebiscite to decide under which country's authority the Rosenthal and the Jauntal in South Carinthia will be placed. The majority of eligible voters voted for Austria and the border which is still valid today was drawn (Klemenčič 2006/07, pp. 29–30). After the First World War, a quarter of Slovenian ethnic territory was placed under Italian authority. After the Second World War, the better part of the territory populated by Slovenians was placed under Yugoslavian authority, while the province of Trieste, the province of Gorizia and Val Canale in the North, together with the rest of the province of Udine – the valleys of Natisone, Torre and Resia – remained a part of Italy (Vouk 2004, pp. 125–133). There had been no mention of a Slovenian minority in Croatia until the dissolution of Yugoslavia as Slovenians were one of the constituent nations of the former country. There are indigenous settlements of Slovenians as well as immigrants from Slovenia in Croatia. In practice, there is no difference between the two categories as they are interwoven. Slovenians live in Gorski Kotar, north Istria, Medžimurje, the Sotla Valley, the Kolpa Valley and in numerous cities (Zagreb, Rijeka, Varaždin, Poreč, Umag, Buje, Čakovac, Split...) (Žagar 1995, pp. 347–348).

3.1 The organization of the Slovenian National Community in Austria

Carinthian Slovenians have not yet overcome the political division from the times after the Second World War up to this day. At that time Yugoslavia supported only the left-wing Union of Slovenian Organizations (USO) and was opposed to the foundation of the democratic National Council of Carinthian Slovenes (NCCS) and did not acknowledge it until 1965. The organization was in a clearly underprivileged position up until Slovenia's attainment of independence. Today the USO still holds diametrically opposite views on two essential items of the program of the NCCS: the idea of having a democratically elected joint representation, which the NCCS strongly advocates, but the USO does not approve of, and the second the dilemma about whether the minority should take part in elections with its own ethnic party or should it join Austrian national parties. The NCCS is an ardent advocate of independent participation and is closely connected with the Unity List (UL), the Slovenian ethnic party in Carinthia. The USO has stressed from the very beginning that it wants to be a unifying organization, but it actually united mostly left-wing political circles and relied mainly on the ideas of the anti-Nazi and Partisan movements. Since its beginnings to this day, the organization has refused independent political participation and advocated the integration into majority parties, especially into the Social Democratic Party of Austria (SPO). After Slovenia gained its independence, the USO lost much of its traditional influence and tried at first to substitute it with increased activity in the Advisory Council for the National Communities, which was organized in the office of the Chancellor of Austria. The USO is today also involved with the Consensus Group which unites some Slovene organizations and some extreme anti-Slovene organizations with the aim of improving their relations. This has also increased the dependence of the USO on the Austrian authorities. They are opposed to the common representation of Carinthian Slovenes and advocate the standpoint that different views within a minority are natural and that there is no need to try to unify them. In their opinion, the only common standpoints that should be adopted within the minority are those regarding the matters which are of vital importance for the minority (Valentinčič 2012, pp. 64–66).

In addition to two traditional organisations, a third organization was founded in 2003, which gradually gained (or is still gaining) the status of an umbrella organization. The Community of Carinthian Slovenian Men and Women (CCS) was created as a consequence of a major dispute within the NCCS, which led to the resignation of the president of the NCCS Bernard Sadovnik. He then established a new organization – CCS – together with a group of like-minded people. The CCS wants to be seen as a non-ideological organization of the future which is interested only in the struggle for the rights of Carinthian Slovenians and in the fostering of the intercultural dialogue (Skupnost – SKS 2013 Web).

Apart from politics, the traditional political division among Carinthian Slovenes can be also seen in cultural organisation. The Slovenian Cultural Union (SCU), a member of the USO, unites the left-wing cultural societies. Today, the SCU unites forty-four societies. On the other hand, the Christian Cultural Union (CCU) unites the societies which identify themselves with the Christian Democratic tradition. In comparison to the SCU and the USO, the relations between the CCU and the NCCS are different. The CCU is not a member of the NCCS, the two organizations function independently and are not formally connected, they only share similar ideological foundations. Today, the CCU unites fifty-three societies and groups (Slovenska kulturna društva na Koroškem, 2013 Web). This duality is a result of the political process and could not be done away with without negative consequences. The incentives to do so appeared after Slovenia gained its independence, but they later died away, or were at least not as strong as with the unifying of political organizations. Malle (1997, p. 5) thinks that without the two organizations, "the activity of the local societies would soon diminish and completely die away with time". For a long time, the two cultural organizations were rivals and did not cooperate, but in recent years they have been connecting more and more. They also prepare joint projects and divide fields of work. Approximately twenty-five local societies are today members of both umbrella organisations (Lesjak Klun 2010, p. 12).

Besides Carinthia a small Slovenian ethnic community also lives in Styria. The only organization they have is Cultural Society Article 7 for Austrian Styria. The society was founded in the late 1980s. Their activities are focused on cultural work, there is no political initiatives. They also own a house, called Paulus house, where they have their seat and organize events.

3.2 The Organization of the Slovenian Linguistic Minority in Italy

Slovenian organizations in Italy were abolished and prohibited when the fascists came to power. They were reorganised following the Second World War and defined the border between the two countries. The same as in Carinthia, the Slovenian minority in Italy is politically divided and therefore separately organized. The left wing of the minority represents the Slovenian Cultural and Economic Association (SCEA), which was founded with the support of Yugoslavia in 1954. It was established by the existing cultural associations which remained faithful to Yugoslavia also in the period following the time of the Informbiro. Sergij Pahor (2010, p. 9) estimates that the individual associations played a very important role for the minority but had little political power in general. Until the collapse of Yugoslavia the SCEA had a dominant position within the minority especially in the economic field, since it had received strong support from Yugoslavia. The other ideological side was solely dependent on voluntary work. Therefore, the SCEA could develop significantly stronger structures in the province of Udine. Until 2012, only organizations were allowed to become members of the SCEA, but now individuals can also join (Zgodovina 2013). In terms of politics, the SCEA strongly opposes the idea of functioning in an independent ethnic party and advocates the integration of Slovenians in the Italian left-wing parties. The SCEA coordinates the activities of more than 160 affiliated associations, organizations and institutions. In Italy, the separation of political and cultural umbrella organizations is not as clear as in Carinthia. There are two umbrella organisations that bring together all minority structures and represent the minority at all levels. Cultural organizations are organized in associations which are also the members of the umbrella organizations (Članstvo 2013 Web).

Slovenian associations in Italy who did not accept the Communist ideology established the Council of Slovenian Organisations (CSO) in 1976. Compared with the SCEA, the CSO was in an subordinate position because it was not funded by Yugoslavia. Therefore, it could only play a representative role (Pahor 2010, p. 9). The situation in Italy is similar to the Carinthian one with more and more associations joining both umbrella organisations. The CSO is closely related to the Slovenian ethnic party in Italy called Slovenian Community (SC). The organization has traditionally been stronger in the province of Gorizia with more conservative rural communities. In the province of Trieste the structures of the CSO were established by the refugees from Yugoslavia who settled in Trieste after the Second World War. Compared with the SCEA, the CSO is in the minority in the province of Trieste. In the province of Udine its structures began to develop after Slovenia gained independence and the CSO began to receive financial support from the newly formed Slovenian state.

The Slovenian cultural organizations in Italy are organized in different associations functioning as a kind of cultural umbrella organizations. However, they were also the founding members of both political umbrella organizations. The two political organizations are more important than the cultural and also have the most significant role in the cultural representation of the minority. The left oriented cultural organizations are organized in the Association of the Slovenian Cultural Organisations. It was founded in 1945. Today it includes 87 member organizations and groups from all three provinces of the Friuli-Venezia Giulia region where Slovenians live. The organizations based on Christian and democratic principles did not want to affiliate to this association. They do not have their representative association. The organizations from the province of Gorizia founded the Slovenian Catholic Union of Education in 1959. Today it has 12 member organizations. Similar organizations from the province of Trieste founded the Slovenian Education Union in 1948. It includes 31 cultural organizations (Tomšič 2010, p. 13). In province of Udine Association Evgenij Blankin connects Slovenian organizations based on Christian and democratic values.

3.3 The Minority Organizations of the Slovenians in Hungary

Until the democratisation of Hungary, the Slovenians were organized with other South Slavs in the Democratic Alliance of South Slavs, which stopped working after the democratisation. All then existing cultural groups and associations merged into the Association of Slovenians in Hungary. Today, the Association includes all 16 Slovenian associations in Hungary. Its main commitment is to improve the situation of the Slovenian minority in the Raba region, in particular to coordinate the activities of various organizations, and to establish a dialogue with the authorities (Munda Hirnök 1999, p. 3). Cultural activity of the Slovenians in the Raba region is exclusively amateur. Although the Association brings together cultural associations it is also an umbrella organization in the field of politics, since the Slovenians in Hungary do not have any political organizations in the true sense of the word.

Hungary provides the so-called self-governments for political and administrative representation of minorities. Self-governing communities are funded by the state and are the direct interlocutor of the Hungarian government. That gives them great importance. The Slovenians are organized in the local minority self-governments in all seven villages of the Raba region settled by the Slovenian minority and in the cities of Szombathely, Budapest, Mosonmagyarovaru and

Szekesfehervaru. These local self-governments form the general Slovenian self-government which is headquartered in Felsőszölnök and also has a representative office in Budapest (Just 2009, p. 8). The self-government is a political organization that closely cooperates with the Association of Slovenians in Hungary since both organizations fight for the same goals. (Ibid).

3.4 Organization of Slovenians in Croatia

Until the dissolution of Yugoslavia Slovenians were one of the constituent nations of the country and there was no mention of Slovenian minorities within the republics. Later, Croatia acknowledged their minority status and several Slovenian societies were founded as well. Today, there are sixteen Slovenian societies in Croatia. It is the Association of Slovenes in Croatia which unites them and is also a political representation of the minority of some sort. The Slovenian minority is not able to have a proper political representation and everything is organized by the societies (Kdo smo? 2013 Web).

Croatian legislation provides also the councils of national minorities, which supervise the practical application of legislation. The councils of the Slovenian national minority were elected on the levels of counties, municipalities and cities. In a county where 100 Slovenes live, they have the right to elect their representative in the council, whereas in the counties where there are at least 200 Slovenes, they can elect their own council. So far, the Slovenian minority has never taken full advantage of its rights regarding the elections of the councils.

The Slovenians in the provinces of Trieste and Gorizia, and in Carinthia are therefore much better organized in terms of politics and culture than the Slovenians in the province of Udine, in Styria, Hungary and in Croatia. This is due to their larger number and larger scope of liberties that they enjoyed in the past. To become politically active people have to develop a strong national consciousness. Minority education, the media and the church have a great influence on it. This is the basis for developing cultural activities that strengthen the national consciousness. The lack of those elements leads to poor political organization in these areas and consequently to a lower level of rights and much higher assimilation.

3.5 Minority School Systems in the Neighbouring Four Countries

Besides political and cultural organizations Slovenian schooling is of key importance for the perseverance of Slovenian language, culture and identity. Slovenians in Italy, in provinces of Trieste and Gorizia have schools, where all subjects are in the Slovenian language, only Italian as a language is obligatory subject. In the province of Udine only one bilingual school functions, all other schools are in Italian only (Valentinčič 2012, pp. 151–159). In Carinthia, besides German also bilingual schools are there, where both languages are supposed to be equal. In Styria only some schools offer the Slovenian language as an elective course, the rest of the school is in German only (Id, pp. 83–86). In Hungary couple of schools offers the Slovenian language as an elective course, besides two elementary schools that are bilingual (Id, 191–194). In Croatia the Slovenian language is mostly offered as a free Saturdays lessons, which Slovenia provides teachers for. Never-the less in the last years the Slovenian language started to be offered as an elective course also in public schooling in Varaždin and Rijeka (Id, pp. 223–227).

4. A Short History of Emigration from Slovenian Ethnic Territory and Slovenian diaspora communities

There were three major waves of emigration from the Slovenian ethnic territory. The first one was from the 1880's[1] till WW1. At that time the reasons for emigration were mostly economic.[2] The country of immigration was mostly the USA, to some extent also South America, Western Europe and Egypt[3]. The second wave was between both world wars. At that time most of the migrants were still the economic ones, but there were also a lot of political refugees (especially from the part of Slovenian territory that became part of Italy and in fever numbers also from the territory that became part of Austria). Migrants moved mostly to South America (especially Argentina), to some extent also to Western Europe and Egypt. The third wave started right after the end of WW2 with the political refugees from the communist regime and continued in the next three decades, when political and economic reasons for migration were interlaced. At the beginning the migration was to USA, Canada, Australia and Argentina (in smaller number also to other South American countries), later on many people moved to Western European countries, especially Germany. In the time of both Yugoslavias Slovenians were constantly moving also to other republics of the common country (Drnovšek 1991, pp. 205–208 and Valentinčič 2013, pp. 327–328).

The realities of the Slovenian communities abroad are so different that any generalization is very difficult. Slovenians living in Europe had and have the possibility to return home for a visit more often and consequently it was easier for them and their children to keep alive connections with the country of origin. Today most members and especially the young remained in Slovenian societies in Germany.

1 The first emigration started about 20 years before that.
2 In some cases also young men excaped from the military draft.
3 So called Aleksandrinke were moving there.

They are organized in Slovenian societies and catholic missions (they do not own the churches, but are guests in German churches) (Valentinčič 2014).

The Slovenian post-WW2 community in Argentina is regarded to be the symbol of Slovenian identity and language. Besides the fact that whole families moved there and the percentage of ethnically mixed marriages was low, it was very important that many intellectuals were part of this migration who had above average abilities to organize the community (Žigon 2001, pp. 249–257). As an example, one proof of the intellectual power of the community is that less than 7.000 people have been issuing 74 different newspapers. In the year 1957 30 of them at the same time (Pertot 1991). They own various Slovenian national homes. The catholic church and clergy play an extremely important role (Valentinčič 2012/2013 and Valentinčič 2013b).

All Slovenian communities in the anglosaxon countries (USA, Canada, Australia) have the same characteristic that the Slovenian language is disappearing very quickly. But if we can say that in the cases of traditional Slovenian minorities in the neighbouring countries the language is the condition for identifying as a Slovenian the situation in the diaspora communities is not the same. The language is the most difficult to be preserved from all ethnic attributes. Also after losing the language people preserve the traditions, the music, the food[4] and remain very proud of being Slovenians, often with excelent knowledge of Slovenian history and ethnography, in some cases also current politics. In all countries Slovenians are organized in Slovenian societies, national homes and also own their own churches (Genorio 1989 and Slovenian American Times 2015).

5. Empirical Study

We have examined profiles of ninety young people, 10 from each country where larger Slovenian communities live, with the following six criteria:

- The number of posts in total, the percentage of posts in Slovenian language, the percentage of posts in the dominant language and the percentage of bilingual posts;
- The same criteria for comments under posts;
- The number of sharings of Slovenian music;
- The number of posts about Slovenian sport teams and individuals;

4 We can even say that the food is the most powerful ethnic identificator.

- The number of posts regarding the referendum on gay marriage and adoption of children into homosexual families, that took place in Slovenia on the 20th of December 2015;
- The number of posts inviting or reporting about Slovenian ethnic events in the communities where they live.

Table 1: Language of post on Facebook

Country	Number of people taken into considearation	Number of all posts	Number of Slovenian posts	Percentage of Slovenian posts	Number of posts in dominant language	Percentage of posts in dominant language	Number of bilingual posts	Percentage of bilingual posts
Austria	10	74	20	27 %	46	62 %	8	11 %
Croatia	10	77	6	8 %	71	92 %	0	0 %
Hungary	10	118	28	24 %	74	63 %	16	13 %
Italy	10	319	218	68 %	65	20 %	36	12 %
Argentina	10	400	166	42 %	197	49 %	37	9 %
Australia	10	264	7	3 %	257	97 %	0	0 %
Canada	10	362	13	4 %	340	93 %	9	3 %
Germany	10	232	150	65 %	74	32 %	8	3 %
USA	10	473	3	1 %	467	98 %	3	1 %

Results regarding the frequency of use of Slovenian language among Slovenian communities are not suprising at all. Slovenians in the provinces of Trieste and Gorizia in Italy have the best political and cultural organizations, have compact area of population and have schools in the Slovenian language. They are considered to have a high level of national awareness. All these factors result in a high percentage of Slovenian posts. Italian posts are mosty consequence of sharing Italian news or trying to inform Italian friends about something, where as bilingual posts are about information or opinions that are, in the opinion of the authors, relevant for Slovenians and Italians as well. The fact that they do not decide for posts in Italian only (which is understood by Slovenians from Italy) shows the strength for perserving the language and identity. Slovenians in Austria are still well politicaly and culturaly organized, but their area of the population is more spread and mixed with the German speaking population. Elementary schools are only bilingual and limited percentage of population continues their schooling in Slovenian/bilingual high schools. They have more daily contacts and friends among dominant ethnic group and consequently mosty use the German language, only occasionally Slovenian. It is well known fact that they are not considered to be as nationaly aware and courageous, some even claim that they have an inferiority complex towards the German speaking population. All this is reflected on Facebook too.

We can say that resoults for two countries are a bit surprising. The first case is Hungary. Slovenians in the Raba region do not have as so strong organizations as those in Austria and Italy do, their schools system is much poorer too. School advisor from Slovenian Valeria Perger constantly claims on her blog[5] that the situation is so bad that Slovenian is not a language used in every day communication any more and it is becoming just a folklore. Never-the-less the percentage of posts in the Slovenian language is almost as high as it is in Austria. The most logical explanation for this is that little more than one year ago a society of youth from the Raba region was organized.[6] Since the aim of the organization is to preserve Slovenian traditions, they try to be very much connected with organizations in Slovenia and they receive funding from Slovenia, we can conclude that it apparently it is unformally desirable to use the Slovenian language regarding their activities. Since a lot of activities were going on in the time of our study quite some Slovenian posts appeared.

It is not surprising that the lowest percentage of use of the Slovenian language is among young Slovenians from Croatia. A small number of people, the Slovenian language is just a course and not in shools[7], poor cultural organization and almost no political organization at all result in very high assimilation. Those young people that posted thinks in Slovenian (too) were connected to Slovenian because they decided to attend high school or university there.

The second surprise is the high percentage of posts in the Slovenian language among young Slovenians in Germany. There is no doubt that this population is more closely connected with Slovenia because of the physical vicinity (and also not forgeting the fact that the majority of emigration there took place later than to over-sea countries). Nevertheless, living in the German enviroment and have almost two thirds of posts in Slovenian language shows, that apparently they have

5 Iskriceizporabja.blogspot.com. The blog is raising a lot of complaints and resentment.

6 This is actually an organization gethering young Slovenians, but does not use this term in its name, because of the complex identity of those people.

7 With some eceptions.

much more Slovenian than German friends and are also more important for them. This even shows some sights a ghetto situation.

It is not suprising that the percentage of Slovenian posts in Australia, Canada and USA is very low, the lowest in USA. As mentioned above, in these communities the Slovenian language is today the less important fact of ethnic identification. The situation is quite different in Argentina, where we can see that half of posts are Slovenian or bilingual, four fifths of them Slovenian. The Slovenian community in Argentina has always given great importance to the maintenance of the language and the third generation still respects this. Many of these posts are related to the events in the Slovenian community. The fact that language is an imporatant identificator among Slovenians in Argentina is well seen on Facebook too.

Table 2: Language of comments on Facebook

Country	Number of people taken into considearation	Number of all comments	Number of Slovenian comments	Percentage of Slovenian comments	Number of comments in dominant language	Percentage of comments in dominant language	Number of bilingual comments	Percentage of bilingual comments
Austria	10	94	72	77 %	19	20 %	3	3 %
Croatia	10	283	80	28 %	203	72 %	0	0 %
Hungary	10	212	27	13 %	185	87 %	0	0 %
Italy	10	238	181	76 %	54	23 %	3	1 %
Argentina	10	603	165	27 %	438	73 %	0	0 %
Australia	10	401	21	5 %	380	95 %	0	0 %
Canada	10	1042	100	10 %	939	90 %	3	0 %
Germany	10	240	101	42 %	138	58 %	0	0 %
USA	10	1200	60	5 %	1140	95 %	0	0 %

In the case of the language used in comments under Facebook posts we can see the same trends as with the posts. The first difference is not surprising to us. The percentage of bilingual comments is in all cases lower than percentage of bilingual posts. The reason for this is quite simple. With posts their authors can try to share information or their opinion with both linguistic groups. In comments, when the discussion started one language is usually chosen, people do not see the point in dubling what they write in both languages. Often the language of comments depends from the first comment. Although in some cases the discussion is taking place in both languages under the same posts.

The higher percentage of comments in Slovenian, in comparison to the percentage of posts can be interpreted in the context that apparently many young Slovenians try to be not-exclusive – to make sure that the population from the ethnic majority population would understand their posts too. Most of their close Facebook friends are apparently of the same ethnic group, so in the comments the language prevails towards their common language. We see three exceptions here: Hungary, Germany and Argentina. We have previously labeled results for Germany and Hungary as a bit suprising. We can say that it is showen in the comments that the Slovenian language is not that common in their communities anymore, so when the disccusion starts it is easier for them to switch to the dominant language.

Table 3: Posts regarding topics relevant for Slovenian identity

Country	Number of people taken into consideration	Number of all posts	Number of posts regarding Slovenian sports	Percentage of posts regarding Slovenian sports	Number of posts regarding referendum on same-sex marriages	Percentage of posts regarding Slovenian politics	Number of posts with Slovenian music	Percentage of posts with Slovenian music	Number of posts about events in their community	Percentage of posts about events in their community
Austria	10	74	0	0 %	9	12 %	0	0 %	12	16 %
Croatia	10	77	0	0 %	0	0 %	0	0 %	0	0 %
Hungary	10	118	1	1 %	0	0 %	3	3 %	15	13 %
Italy	10	319	11	3 %	11	3 %	2	0 %	77	24 %
Argentina	10	400	0	0 %	10	2 %	1	0 %	33	8 %
Australia	10	264	0	0 %	2	1 %	2	1 %	5	2 %
Canada	10	362	5	2 %	3	1 %	3	1 %	13	4 %
Germany	10	232	10	4 %	0	0 %	18	8 %	25	11 %
USA	10	473	1	0 %	0	0 %	3	1 %	2	0 %

Interpreting all these examples we can find out that Facebook is just a reflection of realities among Slovenian minority and diaspora communities, and not a creator of new realities. The sum of all relevant posts is low, we can see that Facebook is first of all personal media and consequently a reflection of things that occupy people in everyday life. Being Slovenian is not a key element. Most posts are on the topics of events in the Slovenian community – forecasting and reporting about events. The sharing of these posts is comparable with the situation in the communities – the number of events that those communities have and the level of assimilation of Slovenians in those countries. It is completely understandable why Italy, Austria, Germany and Argentina are on the top of the scale. Reasons for Hungary being so high are probably the same as described before. We would expect a bit higher percentage in Argentina. Maybe the reason was that seasons are opposite there, so maybe not that many events took place. On the other hand they have so many constant events that maybe this is a regularity there and they do not even inform about events anymore.

There is substantially less posts regarding all other criteria (we are interpreting the question of same-sex marriages and adoptions below, so we are living this out now). Slovenian sports are very popular among Slovenians in Italy and Germany. Both communities are known as tightly connected to Slovenia, so this is not suprising to us. The cases of USA and Hungary are just exceptions. The reason for more posts on this topic in Canada (in comparison to comparable countiries) is not easily explanable.

In the case of Slovenian music a very clear distinction between minority and diaspora communities are seen. The difference is well known – in diaspora communities the traditional Slovenian music is a much more important identification element. Young people still do listen to it and identify with it. In the neighbouring countries on the other hand people follow the modern music trends. Since they have more connections with the popular culture of countries where they live than the Slovenian one, Slovenian music is not that often popular among them anymore.

Table 4: Posts regarding same-sex marriages and adoptions – the porportion between pro and contra posts

Country	Number of people taken info consideration	Number of all posts	Number of posts supporting same-sex marriages	Percentage of posts supporting same-sex marriages	Number of posts objecting same-sex marriages	Percentage of posts objecting same sex marriages
Austria	10	9	9	100 %	0	0 %
Croatia	10	0	0	0 %	0	0 %
Hungary	10	0	0	0 %	0	0 %
Italy	10	11	6	55 %	5	45 %
Argentina	10	10	0	0 %	10	100 %
Australia	10	2	0	0 %	2	100 %
Canada	10	3	3	100 %	0	0 %
Germany	10	0	0	0 %	0	0 %
USA	10	0	0	0 %	0	0 %

We can see that the number of posts here is very low, so it is very difficult to conclude with relevant findings that can be generalised. There is no doubt that the interest for politics in the mother country is very indicative regarding the level of national awareness and connection with the mother nation. For Slovenians the question on gay marriages and children adoptions was the key political question in 2015, that raised most feedback on social media. So it was understandable to choose this question for our analysis. Since this question is a global issue in today's world we would expect that also people who are not so closely connected with Slovenia would find this question intriguing. Taking into consideration the number of all posts we can see that obviously that was not the case.

Regarding pro and contra, results are not surprising at all for Argentina. It is common knowledge that there political emigration is very much interested in Slovenian politics and also keeps strong catholic belief. Any other result than activation against this changes would be a surprise.

The fact that there were not any posts regarding this topic in Hungary, Croatia and USA is not a surprise ether. The level of assimilation of Slovenian communities in Hungary and Croatia is relatively high, whereas in the USA the identity is mostly cultural, not that much political. We would expect different results for

Germany, where Slovenian population shows tight connections with Slovenia on all other issues. Here apparently nobody saw the question as so relevant to activate himself.

The difference between Australia and Canada can only be explained via personal opinions of individuals (who were somehow exceptions) – Since the structure of both emigrants (if we generalize a bit) is very similar and can also not find differences worth mentioning on this topic in the dominant society. We can not consider the fact that the number of posts was low. The reason for this is the same as with the USA, the level of assimilation is getting higher and higher and the Slovenian identity is more and more cultural and becoming less political.

The results are somehow surprising regarding Italy and Austria. There is no doubt that Slovenians in Italy are more closely connected with Slovenians than those in Austria. Nevertheless results on this topic are opposite. Since it is considered that the majority of Slovenians in Italy are polically left-wing orientted, where as the majority of Slovenians in Austria are supposed to be conservative we can find surprising only the proportion of pro and contra posts. The most logical explanation would be the high level of activation in the pro direction by the Club of Carinthian Slovenian Students in Vienna. Appareantly it was socially not welcome for young people from Austria to be against this act.

6. Conclusion

The results of the empirical study clearly show that young Slovenians from all nine countries use Facebook also as a tool for expressing their ethnic identity. This is most often the case in Italy and Germany and the least often in USA, Australia and Croatia. It is clear that the share of "Slovenian" posts is inversely proportional with the level of assimilation in these communities. Regarding the results we can conclude that the following conditions play the most significant role with activation of the young people expressing their identity on Facebook:

- The physical distance from Slovenia and consequently live connections with the mother country (it is often said that with the emergence of the internet physical distance does not play any role anymore, but we can see that this is not so);
- The quality of political and cultural organization of communities, the level of rights that they have and the density of their population;
- The presence of Slovenian language in schools.

With this research we were able to find some general conclusions comparing the situations in nine countries. Chances for the continuation of the research are large. Especially the following would offer more detailed results:

- Including larger number of profiles to be examined;
- Conducting semi-structured interviews with (some) of the Facebook users, whose profiles were examined. This would give us the explanations of the users as to what are their motivations for posting specific things related with their Slovenian identity on their Facebook's and how do they decide which language to use;
- Examinations of (chosen) Facebook groups. People gather in these groups on the principle of similar values and interests (Cheung, Chiu and Lee 2011). It would be very relevant and interesting to compare groups from individual countries (maybe, for the beginning the group that is understood as Slovenian ethnic group and has the most participants for each country).

These researches are especially relevant since we can expect that the importance of social media regarding the maintenance and development of ethnic identities will constantly rise, especially in the cases of minority and diaspora ethnic groups. We can say that today Facebook is still just a reflection of realities among the Slovenian minority and diaspora communities, and not a creator of new realities. That might change very soon in the future.

7. References

Boykova, Lyudmila: *Online social activism of students: social media as an innovative tool for personal empowerment.* Doctoral dissertation. School of Advanced Social Studies: Nova Gorica 2015.

Cheung, Christy / Chiu, Piu-Yee / Lee, Matthew: Online social networks: Why do students use facebook?. *Computers in Human Behavior*, n. 27, 2011, p. 1337–1343, retrieved 18.12.2015, from: http://wiki.stoa.usp.br/images/e/e6/Why facebook.pdf.

Drnovšek, Marjan: *Pot slovenskih izseljencev na tuje.* Madika: Ljubljana 1991.

Elkins, David: Globalization, Telecommunication, and Virtual Ethnic Communities. *International Political Science Review*, year 18, n. 2, 1997, p. 139–152.

Genorio, Rado: *Slovenci v Kanadi.* Inštitut za geografijo: Ljubljana 1989.

Jones, Steve: *Doing Internet research: Critical Issues and Methodes for Examining the Net.* SAGE: London 1999.

Just, Franc: *Porabje.* Franc-Franc: Murska Sobota 2009.

Kacin Wohinz, Milica / Pirjevec, Jože: *Zgodovina Slovencev v Italiji 1866 – 2000*. Nova revija: Ljubljana 2000.

Klemenčič, Matjaž / Klemenčič, Vladimir: Prizadevanja koroških Slovencev za narodnosti obstoj po drugi svetovni vojni. *Mohorjeva založba: Klagenfurt-Ljubljana-Wien 2006/07.*

Krščanska kulturna zveza: *Slovenska kulturna društva na Koroškem*, retrieved 12.10.2015, from: http://www.kkz.at/drustva_skupine/C16/.

Lesjak Klun, T: SPZ in KKZ sta osrednji kulturni organizaciji, v katerih se združujejo Slovenci na avstrijskem Koroškem. *Novi razgledi*, n. 17–18, 2010.

Malle, Anton: *Organizacije, društva in ustanove slovenske narodne skupnosti na Koroškem. Poskus opredelitve njihove zasidranosti in vpliva med Slovenci na Koroškem*. Slovenski znanstveni inštitut: Klagenfurt 1997.

Medvešek, Mojca: Podoba slovenske narodne manjšine, ki živi na območju avstrijske Koroške, na svetovnem spletu. *Razprave in dokumenti*, n. 7, 2012, p. 8–51.

Medvešek, Mojca: Ugotavljanje podobe slovenske skupnosti, ki živi na območju Furlanije Julijske krajine, na svetovnem spletu. *Razprave in dokumenti*, n. 60, 2009, p. 6–38.

Munda Hirnok, Katalin: *Kratka ocena položaja Slovencev na Madžarskem*. Inštitut za narodnostna vprašanja: Ljubljana 1999.

Newsom, Victoria A. / Lengel, Lara (2012) Arab Women, Social Media, and the Arab Spring: Applying the framework of digital reflexivity to analyze gender and online activism. *Journal of International Women's Studies*, n. 13(5), p. 31–45, retrieved 7.9.2015, from http://vc.bridgew.edu/jiws/vol13/iss5/5.

Pahor, Sergij: Slovenci v Italiji po drugi svetovni vojni. *Novi razgledi*, n. 14, 2010.

Rheingold, Howard: *Virtual Reality: The Revolutionary Technology of Computer-Generated Artificial Worlds - and How It Promises to Transform Society*. Simon & Schuster: New York 1991.

Skupnost koroških Slovencev in Slovenk: *Skupnost – SKS*. Retrieved 9.9.2015, from http://www.skupnost.at/kdo-smo/skupnost-sks/.

Slovenian American times: *Intervju z g. Dejanom Valentinčičem*. Volume VII, Issue 12, 15 October 2015, p. 26.

Slovenska kulturno gospodarska zveza: Članstvo. Retrieved 9.9.2015, from http://www.skgz.org/clanstvo.

Slovenska kulturno gospodarska zveza: *Zgodovina*. Retrieved 9.9.2015, from http://www.skgz.org/zgodovina.

Tomšič: Delovanje Zveze slovenski kulturnih društev. *Novi razgledi*, n. 13, 2010.

Valentinčič, Dejan: Medsebojna soodvisnost pravne zaščite in organiziranosti slovenskih manjšin v sosednjih državah – primerjava med državami. *Dignitas*, n. 55–56, 2012, p. 88–125.

Valentinčič, Dejan: Na obisku med Slovenci v daljni Argentini. In: *Signal*. Kulturno društvo Člen 7 za avstrijsko Štajersko: Graz 2012/2013.

Valentinčič, Dejan: Na obisku med Slovenci v daljni Argentini. In: *Mohorjev koledar 2013*. Mohorjeva založba: Klagenfurt 2013b.

Valentinčič, Dejan: *Primerjava pravnih zaščit in organiziranosti slovenskih manjšin v sosednjih državah*. Master's thesis.Evropska pravna fakulteta: Nova Gorica 2012.

Valentinčič, Dejan: Slovenci iz celotne Nemčije so se zopet srečali. *Slomedia.it*. Retrieved 12.9.2015, from http://www.slomedia.it/slovenci-iz-celotne-nemcije-so-se-zopet-srecali-foto-galerija.

Valentinčič, Dejan: Slovenija in rojaki izven meja matične domovine. *Evropska Slovenija – prispevki za nov nacionalni program*. Nova revija and Inštitut Karantanija: Ljubljana 2013.

Vouk, S: Slovenci v Italiji od mirovne pogodbe do Londonskega sporazuma. In: Gorazd Bajc, ed: *Na oni strani meje*. Univerza na Primorskem, ZRC Koper, Zgodovinsko društvo za južno Primorsko: Koper 2004.

Verhlust, Stefan: Diasporic and transnational communication: technologies, policies and regulation. *Javnost*, year 6, n. 1, 1999, p. 26–36.

Warschauer, Marco / De Florio-Hansen, Inez: Multilingualism, identity, and the Internet. In A. Hu / I. De Florio-Hansen, eds: *Multiple identity and multilingualism*, p. 155–179. Stauffenburg: Tübingen 2003.

Zhao, Shanyang / Sherri, Grasmuck / Jason, Martin: Identity construction on Facebook: Digital empowerment in anchored relationships. *Computers in human behavior*, n. 24 (5), 2008, p. 1816–1836.

Zveza Slovencev na Hrvaškem: *Kdo smo?*. Retrieved 9.9.2015, from http://slovenci.hr/si_ZVEZA/.

Zwitter, Franc: *O slovenskem narodnem vprašanju*. Slovenska matica: Ljubljana 1990.

Žagar, Mitja: Ali so Slovenci na Hrvaškem narodna manjšina: splošna terminološka vprašanja in pravna zaščita. In: Vera Kržišnik-Bukić, ed: *Slovenci v Hrvaški*. Ljubljana: Inštitut za narodnostna vprašanja: Ljubljana 1995.

Žigon, Zvone: Slovenska politična emigracija v Argentini. In: *Slovensko izseljenstvo*. Slovenska izseljenska matica: Ljubljana 2001.